Roman's Notes on DNA

ROMAN B. ROMANIUK

Trifolium Books Inc.

Trifolium Books Inc.
238 Davenport Road, Suite 28
Toronto, Ontario, Canada M5R 1J6

Canadian Cataloguing in Publication Data
Romaniuk, Roman B. (Roman Boris), 1954-
   Roman's notes on DNA

ISBN 1-895579-93-7

1. DNA. 2. Genetic code. 3. Biotechnology. I. Title.

QH430.R65 1996   574.87'3282   C96-931511-2

Printed and bound in Canada
10 9 8 7 6 5 4 3 2 1

Editing: Rosemary Tanner
Production Coordination: Francine Geraci
Cover Illustration: Linda Hendry
Text Design: Jack Steiner

If you would like to know about other Trifolium titles, please visit our Web Site at:
   www.pubcouncil.ca/trifolium

First edition printed in 1995 by DNA SEMINARS™ Toronto, Canada. Copyright ©1995 by Roman B. Romaniuk. ISBN 0-9680280-0-4

Acknowledgments

I suppose every author thinks back to the time when they were first stricken with the inspiration to write a book. In my case, it was after my second year of giving slide show seminars about DNA and biotechnology to biology students at all levels. Time and time again, students would approach to thank me for the neat little memory tricks that I had given them. After almost every seminar I gave, at least one student would ask me to put all these ideas between two covers and make them available in a booklet. They said they would recommend it to their friends and relatives who were curious about DNA because it would make it easier for them to understand the basics. That is what did it for me.

I would like to thank Dr. Lois Edwards for reviewing the text and Professor Don Galbraith and Trudy Rising for their support and belief that I actually have something here!

Many thanks to Steve Hegedus, Roop Misir, and Maxine Widmeyer for their review and recommendation of this book to the publisher, and thanks to Viviane Joly for her assistance in the preparation of the original manuscript.

Special thanks to Michael Shimbos and Danton O'Day, who inspired me in my early academic years to view my studies in biology as a way of looking at the world at large.

Table of Contents

INTRODUCTION

Hi, my name is Roman. In case you're wondering about the cover drawing of the guy gliding along the twisted ladder, that's me having some fun with DNA! I was once a student like you, working my way through a Biology course. If you asked me why I wrote this little book of helpful hints, I'd say it's to give you an edge in your study time so that you can indulge yourself in more activities of the enjoyable outdoors kind. Although Biology has far more to do with the understanding of interrelated concepts than with the mere memorization of a lot of facts and figures, as a student I couldn't help but wonder how much easier my life would be if I had a few good memory tricks up my sleeve that would help me to zip through difficult textbook pages without grinding my teeth. That's what I hope this book will do for you. It is a guide to help you remember some important facts and figures about the characteristics of life, DNA, the genetic code, and biotechnology. Although this book is certainly not meant to replace your textbook, it should give you a little boost in cracking open those chapters dealing with these subjects. I've picked a few areas that are generally hard to grasp and broken them down into bite-sized bits to give you a better understanding of some key concepts that you can build upon. And I've written a Glossary to help you understand the jargon.

And now, let's get to the good stuff! I happen to be fascinated by DNA and the genetic code because they are the basis of the structure and function of all living things on this planet. With a good understanding of DNA and how it serves as the blueprint for all life, one can't help but develop a wonderful appreciation for the incredible diversity of living things and what makes them tick. While DNA is ultimately responsible for the great variety of life on Earth, it is also the common thread that unites all living things, from bacteria to humans. It is the carrier of the genetic code that is universal to all creatures large and small. Although DNA has been around for at least three and a half billion years, only in this century has science begun to unravel its secrets. In the very short time since the mystery of its structure and function has been revealed, many fascinating applications have already been made with this newly acquired knowledge.

These days you can't open a newspaper or turn on your TV set without hearing about DNA and genetic engineering. You see and hear about DNA with increasing frequency because of all the experiments and applications that scientists are performing with it.

To take you back in history a bit, the 1950s and 1960s saw the emergence of sophisticated rocket technology that took us into the Space Age. In the 1970s, the emergence of microchip technology brought us into the Computer Age. Now, in the 1990s, the understanding of genes at the molecular level and the ability of biotechnologists to manipulate the genetic code has thrust us into the DNA Age!

Genetic experiments have evoked responses ranging from awe and optimism for the future survival of humankind to alarm over what some people consider is unnecessary tampering with nature. A great deal of attention will be paid to the DNA molecule in the 21st Century. The better acquainted you are with this molecule, including what it is, where it's found, and what it does, the better you will understand the rapid changes as they take place in the world around you. With this book, I like to think that I am contributing to "DNA literacy."

Because DNA and genetic engineering are topics that many consider the realm of ivory-tower science, most people tend to shy away from them. With this book I hope to melt away this fear by the gentle and gradual promotion of DNA literacy. It's easy to see how you could make a dramatic change for the better in the lives of illiterate people by teaching them how to read and write. In the same way, I feel that teaching people to be DNA-literate will help them cope with what will likely be some great changes. The capacity to splice together DNA from different organisms ranging from bacteria to humans, and making these newly formed combinations work in living cells, opens up possibilities that scientists hadn't considered a scant 30 years ago.

DNA technologies will touch the everyday lives of every person on this planet and will leave many wondering what it's all about. The Human Genome Project is just such an example. It is the mammoth undertaking of finding the precise location and function of all of the 100 000 or so genes that exist within human DNA. Its projected completion date is the year 2005, just around the corner from now. With its completion will come major breakthroughs in medicine and progress towards gene therapies for people suffering from genetic diseases. But, as with any technological breakthrough, it is a double-edged sword. Will the ability to manipulate the genetic code to treat genetic diseases cause scientists to alter human physical traits for other reasons? Will the ability to map disease genes tempt corporations to implement mandatory genetic screening for purposes of health insurance and employment consideration? DNA-literate people must closely scrutinize the social, legal, ethical, and moral issues that will arise from these developing DNA techniques. DNA-literate people will appreciate the great benefits as well as the social impact that this technology will bring in the very near future.

I didn't mean to sound too serious with this little speech, but you can mark my words that over the next 20 years you will bear witness to some amazing changes in society due to developments in DNA technology.

Okay, get ready to glide along; it's time for your first lesson!

Roman Romaniuk

What Is Life?

Before getting into the facts and figures about DNA and biotechnology, you have to first understand the definition of life. What distinguishes living things from non-living things? After all, DNA is the hereditary blueprint of all living things.

A living thing is composed of a lot of organic molecules that contain mainly atoms of **S**ulfur, **P**hosphorus, **O**xygen, **N**itrogen, **C**arbon, and **H**ydrogen. On their own, these organic molecules would be classified as non-living entities. The magic of life occurs when these inanimate molecules react with each other in ways that both release and absorb chemical energy. The sum of all these simultaneously occurring biochemical reactions results in what we can observe and physically measure as being the characteristics of a living organism.

An acronym is a word whose letters stand for the first letter of another word. I like using acronyms because they help you remember facts. **SPONCH** is a good acronym to help you remember the six different elements that are found in living things: sulfur, phosphorus, oxygen, nitrogen, carbon, and hydrogen.

Eight Characteristics of Living Things

A nice little acronym for eight characteristics of a living thing is the word **PHMIGRAM** (pronounced FMIGRAM). Using PHMIGRAM, made up of the first letter of each of the following words, you can easily memorize information about the characteristics of life.

1. **P**hysical and Chemical Distinction from the surrounding environment.
2. **H**omeostasis
3. **M**obility
4. **I**rritability
5. **G**rowth
6. **R**eproduction
7. **A**daptability and Mutation
8. **M**etabolism

1. Physical and Chemical Distinction from the Surrounding Environment

A living thing is physically and chemically distinct from its surroundings. An insect sitting on a rock is made up of different component molecules than the rock itself. A fish swimming in the water also has water molecules inside its body but it is made up of other different types of complex organic molecules such as proteins, unique to living things. Living things are highly organized and complex in their molecular structure. In contrast, non-living matter such as air, rock, clay, metal, sand, and water are made up of comparatively simple compounds.

2. Homeostasis

Living things maintain themselves in a state of equilibrium, sometimes called a steady state. For example, birds and mammals maintain a constant body temperature. Most living things maintain a certain concentration of many chemicals and a constant pH (acidity) in their internal fluids. This balance is maintained by many different types of chemical reactions that extract energy and building blocks from chemicals in their environment, and return wastes to their surroundings.

3. Mobility

A living thing can move from one place to another. A single-celled organism such as an amoeba can move, as do insects, amphibians, reptiles, and mammals. Although most plants are fixed in place by their roots, the stems, leaves, and flower petals are capable of movement.

4. Irritability

Irritability is the ability to respond to a stimulus. The amoeba will move from a colder region to a warmer one. Higher organisms have a nervous system which enables them to sense their surrounding environment and react to any threat to survival.

5. Growth

Every living thing, from single-celled organisms like the amoeba to human beings, grow in volume and size by incorporating into their own make-up the atoms that they ingest as food.

6. Reproduction

All living things reproduce, increasing their numbers.

7. Adaptability and Mutation

Living things have the ability to adapt to their environment through random, spontaneous mutations (changes) in their DNA. If certain mutations make an organism more suited to its environment, they may be passed on to the next generation.

8. Metabolism

Metabolism is the sum total of all the chemical reactions that occur within a living system. This includes construction (anabolism) of structural molecules, and degradation (catabolism) of food molecules for energy.

Now that we've covered these eight characteristics of life included in the acronym PHMIGRAM, you can apply these rules to any object to put it to the test: Is it living or non-living? You will find this list reliable when determining if something is living or not. One important point to remember is that some inanimate (non-living) objects may exhibit one or a few of these eight key characteristics of life but never **all** of them. Let's try out the acronym PHMIGRAM on a car to test if it is a living or a non-living thing.

1. Physical and Chemical Distinction

To some degree, a car is physically and chemically distinct from its surrounding environment. When it is on the highway, the metal, rubber, plastic, gasoline, and other compounds that go into its construction make it distinct from the concrete road that it travels on and the air that it passes through. However, a car does not contain any of the four classes of organic molecules that are always present in living things: proteins, nucleic acids, carbohydrates, and lipids. These molecules are unique in their physical and chemical properties from any other type of organic molecule. In the final analysis a car is not considered to be physically and chemically distinct from its surrounding environment. Therefore, Rule #1 is not fulfilled.

2. Homeostasis

A car engine exhibits homeostasis when its engine maintains a constant temperature and pressure, so Rule #2 is fulfilled.

3. Mobility

A car is certainly capable of movement, so Rule #3 is fulfilled.

4. Irritability

A car is capable of irritability because it responds to the stimulus of the driver stepping on the gas pedal. The car reacts by accelerating its movement. Rule #4 is fulfilled.

5. Growth

A car cannot grow in size from a compact model to a station wagon, so Rule #5 is not fulfilled.

6. Reproduction

A car cannot reproduce. A full-size car can't give birth to baby cars. Rule #6 is not fulfilled.

7. Adaptability and Mutation

A car cannot change with the chance result of becoming more adaptable to the conditions of the surrounding environment, so Rule #7 is not fulfilled.

8. Metabolism

Although a car undergoes degradative chemical reactions in the form of oxidation of gasoline for the release of energy (i.e., catabolism), it cannot perform any constructive chemical reactions to build energy-containing molecules (i.e., anabolism). So Rule #8 is not fulfilled.

As you can see from the car example, non-living things do exhibit some of the eight characteristics of living things, but not all eight of them! Use your imagination and try this eight-rule test on other non-living things.

In the next section, I will take you on a step-by-step introduction to DNA itself.

DNA: The Blueprint of Life

Have you ever climbed a tree? Have you ever smelled a fragrant perfume? Have you ever touched a block of ice and felt how cold it is? Have you ever gazed admiringly at a golden sunset or listened to music that made you want to get up and dance? Have you ever enjoyed the sweet flavor of an ice-cream cone on a hot summer day?

If you answered yes to some or all of the above, then you have experienced the benefit of having DNA! You see, all of those experiences have to do with the five human senses: smell, touch, sight, hearing, and taste. All of these senses and any other kind of physical ability that you possess is ultimately due to the DNA in every one of your cells.

So the question is, how is DNA able to do all of these things? It is because DNA carries the code for the construction of all proteins. The wide range of types of proteins contribute to the functioning of living tissues. Before you can understand how DNA deserves the title of "Blueprint of Life," I first have to talk a little about **proteins**.

Proteins

You may have wondered what proteins really are, what they are made of, and where they come from. You probably remember being told as a child to drink your milk and eat your vegetables along with your ham sandwich because the proteins in them would help make you healthy and strong. So you know that proteins are present in a lot of foods that we eat. That's good, but let's not forget that almost all of the products that we eat weren't always just food on the dinner table, they were at one time living plants and animals!

To gain a better understanding of what proteins are and what they do, we can use an analogy that simplifies the issue. Everyone who drives a car knows it can break down due to engine trouble. A car engine is made up of many different components that are designed to fit and work together to keep the engine running smoothly. If one component becomes defective, the engine experiences abnormal wear and tear and can break down.

Well, things work in organisms the same way, except that in our car engine the vital components are made up chiefly of metal. In living things, the vital working components are made up chiefly of proteins, carbohydrates, nucleic acids, and lipids. Proteins are vital because they are involved in virtually every aspect of what keeps an organism running smoothly.

Protein Types

About eight different categories of proteins are found in various combinations in living systems (from single-celled organisms to multi-celled organisms as complex as a whale or a human.)

1. Transport Proteins

Transport proteins pass ions (electrically charged atoms) and vital molecules such as glucose and amino acids across cell membranes. Hemoglobin, found within red blood cells in vertebrates, transports oxygen from the lungs to different parts of the body.

2. Enzyme Proteins

Enzymes are biological catalysts that mediate the biochemical reactions that occur within a cell. For example, lactase is an enzyme that catalyzes the breakdown of lactose (a milk sugar). Digestive enzymes, such as salivary amylase, trypsin, and lipase, digest our food. Nucleases are enzymes that accelerate the hydrolysis of nucleic acids.

3. Antibody Proteins

Also known as immunoglobulins, antibodies fight bacterial and viral infections.

4. Contractile Proteins

Contractile proteins can change shape rapidly. Actin and myosin are the proteins in muscle tissue that enable animals to move their limbs. The undulations of cilia and flagella are made possible by proteins called tubulin and dynein.

5. Hormone Proteins

Some of the hormones that regulate the function of various organs in an organism are proteins. For example, insulin secreted by the pancreas regulates blood sugar concentration, and growth hormone mediates the growth and development of the organism.

6. Extra (Storage) Proteins

Also known as amino acid depositories, these proteins act as a storage supply of amino acids used for building new proteins. Ovalbumen is the protein of egg white and acts as an amino acid source for the developing embryo of egg laying organisms. Casein is a milk protein that serves as the major source of amino acids for the growth and development of baby mammals. Ferritin, a protein found in bacteria, plant cells, and animal cells, stores iron.

7. Receptor Proteins

Receptor proteins are cell surface proteins that bind with signalling molecules carried in the bloodstream. When the signalling molecule binds to the receptor protein, it triggers a unique response by the cell. For example, when adrenalin binds to receptors on the surface of liver or muscle cells, it triggers a release of glucose for quick energy.

8. Structural Proteins

Structural proteins include alpha keratin, which makes up skin, fingernails, hair, and bird feathers. Elastin is the major component of ligaments, capable of stretching in two dimensions. Tendons and cartilage are made up of collagen which has very high tensile strength. Leather is almost pure collagen.

If we take the first letter of the first word in our above list, we get:

T - E - A - C - H - E - R - S

An easy way to remember the eight protein types found in living organisms and their relationship with DNA is to use this little play on words: The DNA molecule is the blueprint for all TEACHERS!

It is important to point out here that in living systems, there are more than the eight different types of proteins listed here.* However, to be able to recite these eight types quickly will give you a great head start in linking together any news headlines you see or hear concerning the subjects of DNA, proteins, and biotechnology.

The Structure of Proteins

Proteins are large molecules made up of smaller molecules called **amino acids**. There are 20 different amino acids that go into the construction of all proteins. These are shown in alphabetical order on page 9. Each amino acid molecule can be divided into two parts. As you can see in the chart, the part of each amino acid that is shaded contains an arrangement of atoms that is the same for all 20 amino acids (except for a small difference in the amino acid called proline.) It is this region that takes part in the linkage of individual amino acids to form a chain.

* Proteins not in the "TEACHERS" acronym are as follows. Proteins regulate gene expression such as **repressor proteins** that bind to DNA and prevent RNA polymerase from transcribing a gene that codes for some other kind of protein. Other transcription factor proteins bind to the DNA molecule not to repress but to help with transcription by making it easier for RNA polymerase to bind to the promoter region. And poisonous proteins such as snake venoms, bacterial toxins, and plant toxins like ricin serve as defense mechanisms against intrusion by other organisms.

The 20 Amino Acids Found in Proteins (in Alphabetical Order)

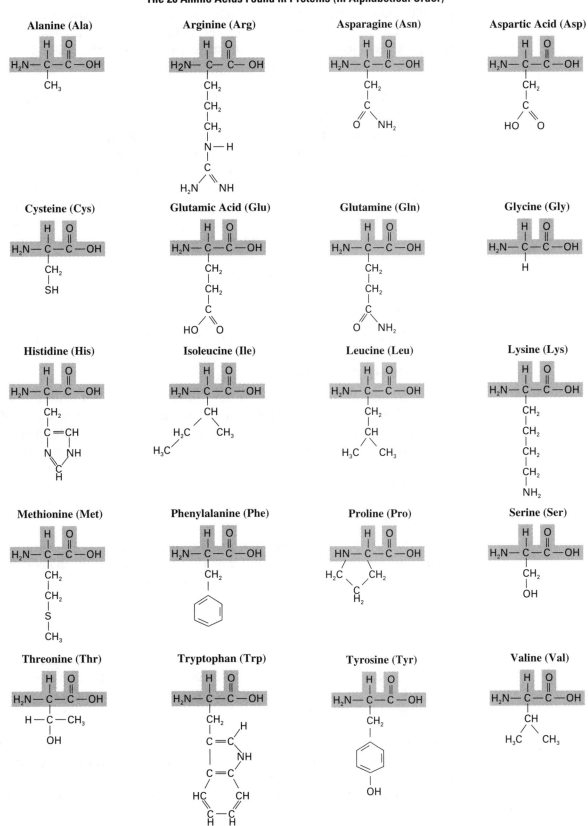

Beneath the shaded section of each amino acid are atoms that make up what are called **side chains**. Each amino acid has a different kind of atomic arrangement in its side chain. It is this unique arrangement of atoms in the side chain that gives each amino acid its own unique electrostatic and chemical properties. This point is an important one to remember when we talk about the 3-dimensional structure of proteins.

The amino acids link up to one another, similar to the links of a chain, by a **peptide bond** that forms between atoms of adjacent amino acids. Since many amino acids are linked together, the chain is called a **polypeptide**. The polypeptide is the first stage in making a final protein. The nice part about this chain formation is that any one of the 20 amino acids can link up to any other one, so you can have an unlimited variety of different amino acid sequences. As we will see, your DNA specifies exactly which amino acids are linked side-by-side to form the polypeptide chains of each protein in your body. This limitless variety of different combinations of amino acids in protein chains is the basis for the diversity of all living things.

There are four levels of protein structure, outlined below.

1. Primary Structure

The sequence of amino acids in the chain (or chains) that make up the protein.

2. Secondary Structure

Hydrogen bonds that form between certain atoms of individual amino acids that are part of the same chain may result in the repeated "coiling" or "folding" of the chain. One example of coiled secondary structure is called an **alpha helix**. This configuration resembles a coiled bedspring, with the side chains attached to the outside of the spring like little spikes. An example of an alpha helix is the secondary structure for alpha keratin, the fibrous structural protein of hair.

3. Tertiary Structure

This is the final 3-dimensional shape of a polypeptide chain. Amino acid chains can fold in a variety of ways to form globular shapes. The twists and contortions that mold the globular shape are due to the electrochemical nature of the side chains of the amino acids.

4. Quaternary Structure

Many large proteins are made up of two or more chains of amino acids, for example, hemoglobin is made up of four polypeptide chains. These chains are held together and shaped by the same kinds of forces of attraction discussed in secondary and tertiary structure.

The most important thing to remember is that the secondary, tertiary, and quaternary structures of a protein are all determined by the primary structure (the order of the amino acids in the chain).

Another important point to remember when discussing the 3-dimensional structure of a protein is that its structure determines the job it does in the organism. Whether it works as an enzyme, a transport molecule, a hormone, a structural support molecule, or an antibody molecule depends on its 3-dimensional structure. As you have already read, the sequence of amino acids ultimately determines the 3-dimensional structure of the protein; thus, the biochemical function is also determined by that sequence. The next big question is:

How Does DNA Determine the Order of Amino Acids in a Protein?

I have outlined how proteins are made up of one or more chains of amino acids. Each chain of amino acids is called a polypeptide, and can be visualized as a string of beads with each bead representing one of the 20 different amino acids. The ordered sequence of amino acids in the polypeptide chain (or to continue to use the analogy, the ordered sequence of beads in a necklace) is dictated by the genetic code contained within the DNA molecule. But first, let's look at DNA.

What Is DNA?

DNA, or **D**eoxyribo**n**ucleic **A**cid, is a long, two-stranded molecule that resembles a twisted ladder, called a **double helix**. As I have said, DNA is found in the cells of all living things, ranging from some viruses and bacteria to trees, grasses, and human beings. It is responsible for passing on hereditary characteristics from one generation of living thing to the next. DNA is called a **nucleic acid** because it is an acidic molecule found in the nucleus of higher organisms. In life forms that contain no nucleus, such as bacteria, the DNA is present in a compact form in the central region of the cell. All living things — whatever the species — from bacteria to humans use DNA as the blueprint for life.

Some scientists discuss whether to classify virus particles as living or not since they are completely inanimate until they infect a host cell. These simplest of all biological entities contain either DNA or RNA as their blueprint for propagation and heredity.

With the help of enzymes, the DNA molecule is capable of **self-replication** (making an exact copy of itself). Its replication is synchronized with the duplication of the entire cell in which it exists during cell division, so that each new daughter cell has the same amount and type of genetic information as the parent cell had.

Each human cell (except the sex cells) contains 46 chromosomes. (Note: sperm and egg cells in humans have only 23 chromosomes each). Each chromosome is made up primarily of one two-stranded thread of DNA wrapped around globules of proteins called **histones**. If you took all 46 chromosomes from a single cell, unraveled each DNA thread, and placed the threads end-to-end, the DNA would be longer than 2 m (6 feet)! This shows you that there is a lot of genetic information packed into the very tiny space of each of your body cells.

To put this into even more astonishing terms, if you took the DNA molecules from all of your body's trillions of cells and attached them end-to-end, the single resulting thread would be long enough to go from the earth to the sun and back about 500 times. Seems impossible, but it's true.

Discovering the Structure of DNA

In 1953, the 3-dimensional structure of DNA was discovered by Dr. James D. Watson and Dr. Francis H. Crick at Cambridge, England. This discovery is one of the greatest breakthroughs in the field of biology in the 20th century. Why? When you know the structure of a molecule, you can then develop insights into its function. Watson and Crick's model for the structure of DNA provided crisp and clear answers to questions of genetics and heredity that had puzzled scientists for more than one hundred years.

DNA is composed of the atoms **P**hosphorus, **O**xygen, **N**itrogen, **C**arbon, and **H**ydrogen (use the acronym **PONCH** to help remember them). The atoms of DNA form the building-block molecules (phosphate, sugar, and nitrogenous base) that are the repetitive components of the overall DNA structure. One of each of these building-block molecules combine to form a **nucleotide**.

Because I feel that it is important that you get a good grasp of the idea of how the two chains of DNA ultimately result in a "twisted-ladder" (or double helix) structure, let's take a step-by-step look at what DNA is made of and how the nucleotides eventually form the double helix.

A Step-by-Step Look at DNA

Each strand of the double-stranded DNA molecule is made up of individual building blocks called **nucleotides** that play a vital role in the genetic code. Each nucleotide in DNA is composed of a **phosphate group**, a **pentose sugar** (sugar with five carbons) called **deoxyribose** (see the illustration below), and a **nitrogenous base**, linked together in that order. Four different kinds of nitrogenous bases go into the construction of a nucleotide: **Guanine**, **Cytosine**, **Adenine**, and **Thymine**. Since there are four different kinds of nitrogenous bases, four different kinds of nucleotides are involved in the construction of DNA.

Each of the two strands of the DNA double helix is actually a **polynucleotide**, a long chain of individual nucleotides attached to one another by chemical bonds between alternating sugar and phosphate molecules. These alternating sugar and phosphate molecules make up what is called the **sugar-phosphate backbone** of

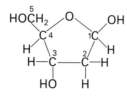

The D in DNA stands for deoxyribose, a 5-carbon sugar molecule. Each carbon atom in the molecule is assigned a number to designate its position in the pentose ring.

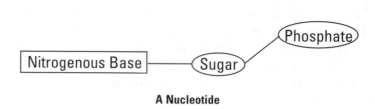

A Nucleotide

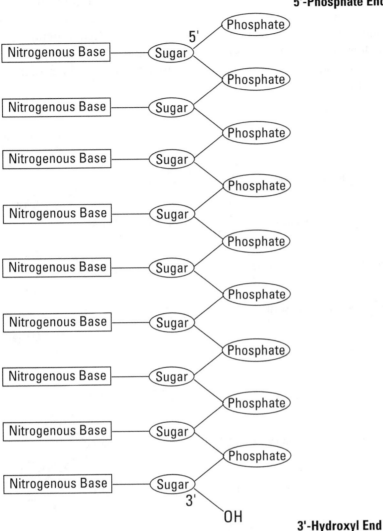

3'-Hydroxyl End

DNA. Here is a simplified illustration of a single polynucleotide chain, showing the sugar-phosphate backbone.

One important feature of the structure of a polynucleotide chain is vital to the understanding of the overall structure of the double helix. The chain has what is called **polarity** with respect to the two ends of the chain. One end has the phosphate group (PO_4) attached to the #5 carbon atom of the deoxyribose sugar, while the other end has a hydroxyl group (OH) attached to the #3 carbon atom of the deoxyribose sugar. Therefore, one end of the chain is called the **5-prime-phosphate end** (5'-phosphate end), while the other end is called the **3-prime-hydroxyl end** (3'-hydroxyl end).

In DNA, two separate polynucleotide chains wrap around each other to form the double helix. I have just mentioned that each polynucleotide chain has a 5'-end and a 3'-end. It so happens that in DNA, the two chains are bound to each other in what is called an **anti-parallel orientation**. This means that while one chain has a 3'-OH end at the top and a 5'-PO$_4$ end at the bottom, the other chain is turned upside-down and has an orientation that is the reverse (i.e., a 5'-PO$_4$ end at the top and 3'-OH end on the bottom.) Here is a simplified drawing of how the two polynucleotide chains of a DNA molecule would look if they were flattened out. The two strands are held together by hydrogen bonds between adjacent nitrogenous bases.

A short piece of the two polynucleotide chains of DNA, flattened out.

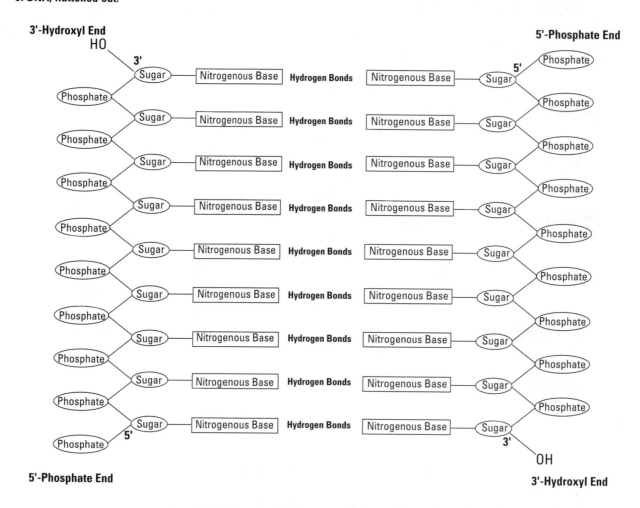

The next figure is a more detailed look of what two unwound polynucleotide chains of a DNA molecule would look like, this time with the actual letters of the nitrogenous base pairs included. Remember, the four nitrogenous bases of DNA are Guanine (G), Cytosine (C), Adenine (A), and Thymine (T). Hydrogen bonds between adjacent bases hold the two polynucleotide chains together.

The DNA in the illustration is still 2-dimensional, with the two polynucleotide chains flattened out on the page to show you the arrangement of the component molecules. In nature, however, molecules assume a 3-dimensional shape that is the most thermodynamically stable (or in other words, most energetically comfortable) for all the atoms that make up the molecule. In the box is the configuration that the atoms in the DNA molecule assume, namely the **double helix**.

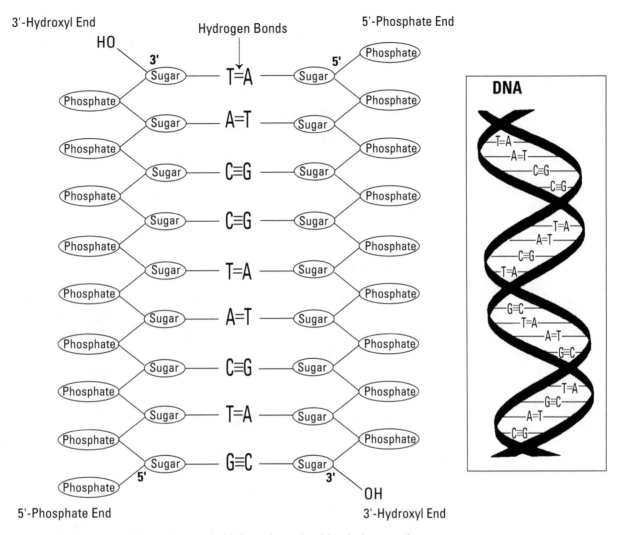

Hydrogen bonds between adjacent bases hold the polynucleotide chains together.

The Difference in Atomic Composition of DNA and Proteins, or PONCH vs. SONCH

If someone asked you to name off the five different types of atoms that make up DNA, you would probably find yourself rolling your eyes from side-to-side, trying to visualize the different atoms in an illustration of DNA you may have seen somewhere in a textbook. If you could name them off, it would probably be with some hesitation.

Well, here is an acronym that will allow you to instantly name the five different atoms that comprise DNA: **PONCH**.

The word **P** **O** **N** **C** **H** stands for:

Phosphorus Oxygen Nitrogen Carbon Hydrogen
These are the five different types of atoms that make up DNA.

The acronym that allows you to instantly name the five different types of atoms that comprise proteins is: **SONCH**.

The word **S** **O** **N** **C** **H** stands for:

Sulfur Oxygen Nitrogen Carbon Hydrogen
These are the five different types of atoms that make up proteins.

In summary then, DNA = PONCH and Proteins = SONCH. These two acronyms very neatly and briefly point out that there is a difference of only one type of atom when comparing the atomic composition of DNA and proteins. The fact that DNA contains phosphorus and no sulfur, while proteins contain sulfur but no phosphorus, is a key point in a famous experiment performed in 1952 by Alfred Hershey and Martha Chase at the Cold Spring Harbor Laboratory on Long Island.

Hershey and Chase knew that bacteriophage (viruses that infect bacteria) contain only DNA and protein. Bacteriophage (or simply "phage") reproduce by injecting their genetic material into bacteria, then use the bacterial systems to direct the production of more phage. Hershey and Chase labeled phage protein with radioactive sulfur (^{35}S) and phage DNA with radioactive phosphorus (^{32}P). After infecting bacteria with the labeled phage, they removed any pieces of the phage from the outer surface of the bacteria by homogenizing the mixture in a blender. Next, they centrifuged the mixture to separate the bacteria from any compounds that were still in solution and not contained inside the bacteria. Their results, summarized in the illustration below, showed that only the ^{32}P, and consequently only the DNA, had entered the bacteria. Since the DNA could direct the production of new phage while inside the bacteria, DNA must be the genetic material.

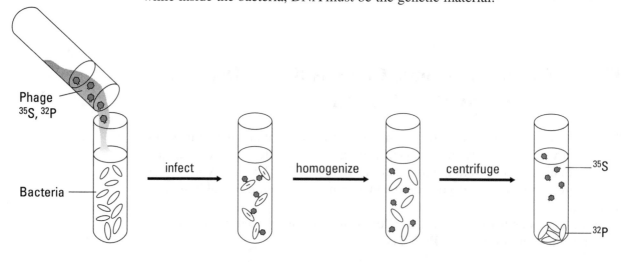

How Do Cells Make Copies of DNA?

Once scientists knew that DNA was the genetic material and what it looked like, they wanted to find out how the DNA was passed on to daughter cells during cell division (mitosis). The DNA had to duplicate itself exactly, so that every cell in the organism had the same DNA, and the same amount of DNA, as in the organism's first cell, the zygote.

Many enzymes and other proteins work together to make two identical copies, or replicas, of DNA from one double helical molecule. This process, called replication, begins when a unique set of proteins binds to a sequence of bases on the DNA (called an origin of replication). One enzyme breaks the hydrogen bonds between base pairs and separates the two strands of DNA as is shown below. Certain proteins bind to the single strands to keep them apart while yet another

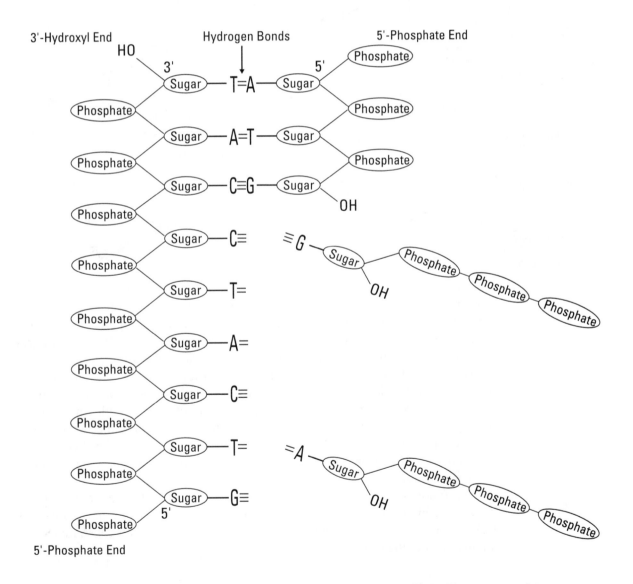

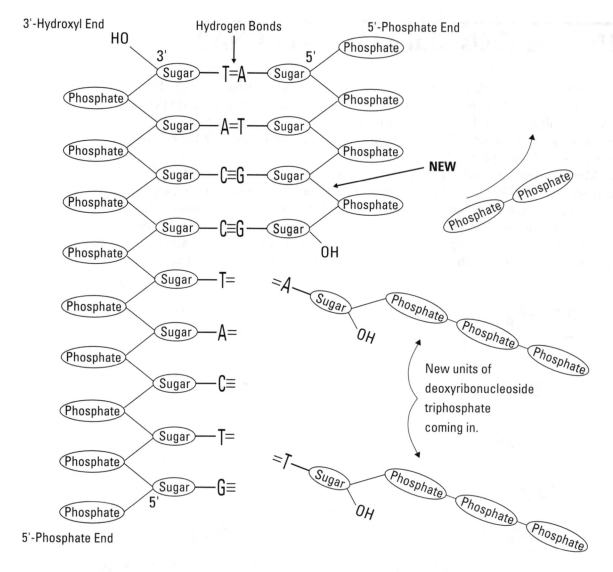

enzyme unwinds the helix further along the chain. In the region where the two DNA stands are separated, the bases are unpaired and can hydrogen bond with any complementary bases that are nearby. This base pairing is the key to ensuring that each new double stranded DNA molecule is identical to the original.

The figure above illustrates one step in replication. Before a new unit attaches to the previously added nucleotide, it has two extra phosphate groups. We call it a deoxyribonucleoside triphosphate. While the correct base is held in place by hydrogen bonds, an enzyme called DNA polymerase catalyzes the formation of the sugar-phosphate bond. During the reaction, the two extra phosphate groups leave and the DNA strand becomes one nucleotide longer. It is now ready to add another new nucleotide in exactly the same way.

The following summary helps you see that each new double-stranded DNA molecule contains one, unaltered, old strand and one completely new strand. As well, you can see how base pairing has ensured that the two new DNA molecules are identical to the original.

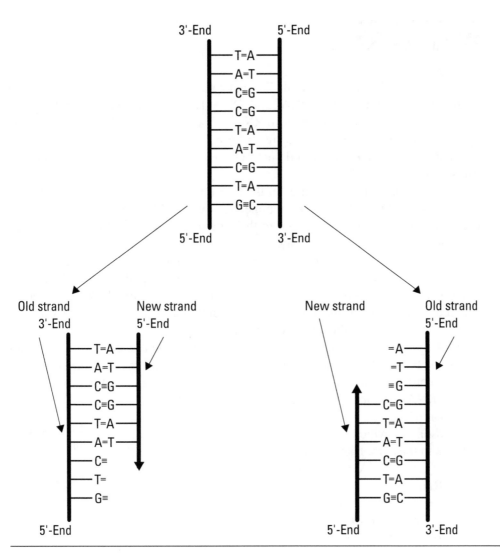

3'-End   5'-End
—T=A—
—A=T—
—C≡G—
—C≡G—
—T=A—
—A=T—
—C≡G—
—T=A—
—G≡C—
5'-End   3'-End

Old strand   New strand
3'-End   5'-End
—T=A—
—A=T—
—C≡G—
—C≡G—
—T=A—
—A=T—
—C≡
—T=
—G≡
5'-End

New strand   Old strand
              5'-End
=A—
=T—
≡G—
—C≡G—
—T=A—
—A=T—
—C≡G—
—T=A—
—G≡C—
5'-End   3'-End

SUMMARY
Two new DNA molecules are formed that are identical to the original DNA molecule.

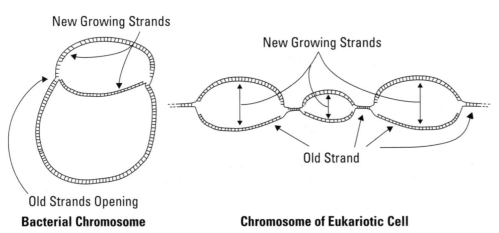

New Growing Strands

Old Strands Opening

Bacterial Chromosome

New Growing Strands

Old Strand

Chromosome of Eukariotic Cell

DNA replication begins at only one place in the circular DNA of bacteria. It proceeds in both directions until the entire circle has been replicated and then the ends are connected.

In the larger chromosomes of higher organisms, replication begins in many different positions along the DNA. It proceeds in both directions from each origin of replication and continues until the entire chromosome has been replicated.

What Is RNA?

RNA stands for **Ri**bo**N**ucleic **A**cid. The name alone reveals one similarity and one difference between RNA and DNA. Like DNA, RNA is a nucleic acid and is made of nucleotides. However, the sugar in RNA is ribose instead of deoxyribose. In the figure shown here, you can see that ribose has one more oxygen atom than deoxyribose (see page 12 if you do not believe me—check the position where the arrow's pointing).

The phosphate group and three of the four nitrogenous bases in RNA are identical to those in DNA. Instead of thymine, however, RNA has a base called **uracil**. The structure of uracil is so similar to thymine that the two bases hydrogen bond with adenine in the very same way.

RNA is single-stranded and consequently does not form one continuous double helix. But many RNA molecules have short sections that are complementary with other short sections in the same strand. Where this occurs, the RNA molecule can fold back on itself and form short segments that resemble the DNA double helix, as shown in the illustration of tRNA.

RNA molecules fit into three categories based on their role in protein synthesis. In one type of RNA, the sequence of bases determines the sequence of amino acids in a protein. Since it carries sequence information from DNA to proteins, this type of RNA is called messenger RNA or mRNA.

Ribosomes, which are small particles in the cytoplasm, contain another type of RNA called ribosomal RNA or rRNA. Bacterial ribosomes have three pieces of rRNA and ribosomes in the cytoplasm of eukaryotic cells have four pieces. (See the Glossary for eukaryotic cells.) In both cell types, the rRNA combines with over 50 different proteins to make up one ribosome.

Transfer RNA (tRNA) molecules form a group of very small RNAs. Each type of tRNA binds to a specific amino acid and brings it to the ribosome where it is added to a growing polypeptide chain. Although each type of tRNA is unique, they all have some common features such as the cloverleaf structure shown here.

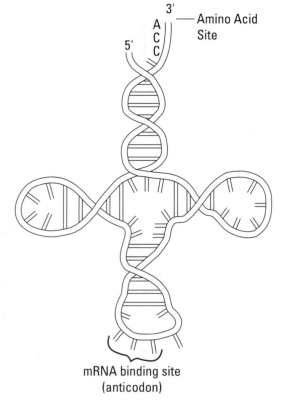

mRNA binding site
(anticodon)

The structure of transfer RNA (tRNA)

How Do Cells Make RNA?

Since RNA and DNA are both nucleic acids, they are produced in much the same way. However, RNA is not identical to DNA, so we do not use the term "replication" to describe RNA synthesis. Instead, we call it "transcription." You can remember this word because to transcribe means to copy down or rewrite in the same language. In the case of DNA and RNA, you can say that the language is "nucleic acids" and the alphabet includes G, C, A, T, and U. Now let's see how transcription differs from replication.

RNA is single stranded because only one side of the DNA is copied. We call the copied side of the DNA the "template strand." The side that is not copied is called the "coding strand" because its sequence is the same as the RNA made from the template strand. Of course, the RNA has a U where the DNA has a T.

Just a small portion of the DNA strand is copied to make one RNA molecule. Hundreds of different RNAs can be made from the single DNA molecule in any one chromosome. Promoters, or unique sequences of bases on the DNA, determine the location where RNA synthesis begins and which side of the DNA will be the template strand in that location. One strand of DNA may be the template strand for a particular RNA, but farther along the same strand may be the coding strand for another piece of RNA.

To begin the synthesis of a strand of RNA, a set of protein "transcription factors" bind to a promoter and begin to separate the strands of DNA. The bases of free ribonucleoside triphosphates form hydrogen bonds with the unpaired bases on the template strand of the DNA. While the nucleotides are held in place, an enzyme called RNA polymerase catalyzes the formation of a sugar-phosphate bond. As the RNA lengthens, the beginning of the strand leaves the DNA which then returns to its double helical form, as shown here. Eventually, a sequence of bases on the DNA signals the termination of RNA synthesis, and the new RNA molecule, along with the transcription factors and RNA polymerase, disconnect from one another. The same section of DNA can be read over and over to make many identical RNA molecules.

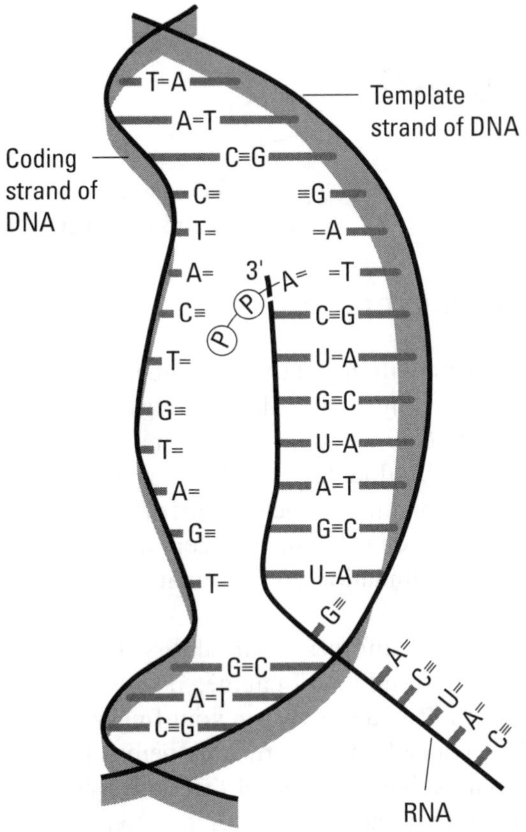

Transcription

Purines and Pyrimidines

The four nitrogenous bases that occur in DNA are Guanine, Cytosine, Adenine, and Thymine. RNA molecules are almost the same except that the Thymine occurring in DNA is replaced by Uracil in RNA.

So: DNA = G, C, A, T (pronounced GeeCAT) and
RNA = G, C, A, U (pronounced GeeCOW)

Purines
(Adenine & Guanine)

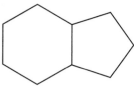

Double-ring structure
(Longer Structure)

Pyrimidines
(Cytosine, Uracil &
Thymine)

Single-ring structure
(Shorter Structure)

Do you ever have trouble remembering which of the nitrogenous bases in DNA and RNA are the **purines** and which are the **pyrimidines**? Worry no more. Here is a simple review of the two types of compounds, and a memory trick you can use to immediately distinguish between the two.

At left are outline drawings of the molecular structures of purines and pyrimidines. For the sake of simplicity they are shown without any of the double bonds that occur between certain atoms in both types of compounds.

Using these as a visual reference, let's see if we can come up with a single word-trick that will bring all this information together at a glance. The word is **PURLAG**. Okay, you might say, what's the key to this word and how does it condense all of the above information?

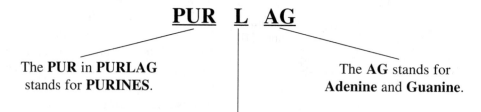

PUR L AG

The **PUR** in **PURLAG** stands for **PURINES**.

The **L** stands for **LONGER STRUCTURES**.

The **AG** stands for **Adenine** and **Guanine**.

By memorizing the word PURLAG, you will instantly remember that the purines are the nitrogenous bases that have the longer (i.e., double-ring) structure and they happen to be Adenine and Guanine. This will come in very handy when you look at textbook illustrations of the DNA double helix because you will instantly be able to identify the double-ring structures as being either Adenine or Guanine. It is obvious that if you know which of the four nitrogenous bases of DNA are the purines, then by the process of elimination you automatically know which ones are the pyrimidines (i.e., C and T).

This next word-association trick further reinforces your ability to know instantly which nitrogenous bases are the pyrimidines. Imagine, if you will, a stainless-steel pyramid with very sharp edges. If you briskly ran your finger along one of those edges, you would likely CUT yourself. The word-trick here is that if you speak out loud, the word **pyramid** sounds like a portion of the word **pyrimidine**. Because you can cut yourself on a pyramid, the word association to remember here is:

Pyrimidine (sounds like Pyramid) = <u>C</u> <u>U</u> <u>T</u> = <u>C</u>ytosine <u>U</u>racil <u>T</u>hymine.

Therefore, you can always associate pyrimidines with the word CUT, the letters of which happen to be the first letters in the words Cytosine, Uracil, and Thymine.

Summary

The four nitrogenous bases in DNA are Guanine, Cytosine, Adenine, and Thymine.

(**GCAT**, pronounced GeeCAT)

The four nitrogenous bases in RNA are Guanine, Cytosine, Adenine, and Uracil.

(**GCAU**, pronounced GeeCOW)

PURLAG means that the **PUR**ines have the **L**onger double-ring structure and they are **A**denine and **G**uanine.

Pyrimidines = **CUT** means that the Pyrimidines (which have the shorter single-ring structure) are **C**ytosine, **U**racil and **T**hymine.

The Genetic Code

In a nutshell, here is an explanation of how the genetic code works. I will discuss protein synthesis in more detail later.

In genetic code language, you only use the first letter of each of the four types of nitrogenous bases. So, the four letters of the DNA genetic code are **G, C, A,** and **T**. Each DNA code-word is made up of any combination of three of these four letters. So TAC, CAT, GAG, TAG, and ACT are just five examples of the 64 three-letter code-words (**nucleotide triplets**) found along one of the two DNA strands that ultimately code for the placement of amino acids into a growing protein chain. This is why DNA is called the **blueprint for all proteins**.

DNA is the master molecule of the genetic code. Its genetic code is passed on to an intermediate courier called the **messenger RNA molecule (mRNA)** which takes an imprint of the DNA code. The mRNA then goes to a protein-making factory in the cell where proteins are constructed one amino acid at a time. The nucleotide triplets of the messenger RNA, called **codons**, are read from one end of the molecule to the other like a ticker-tape.

In the following example, I have used the nucleotide triplets mentioned above. They might be found along one of the two strands of a DNA molecule. To give you a better idea of the flow of genetic information from the DNA molecule, to the messenger RNA molecule, to the placement of amino acids to make a protein chain, here is a three-line description of that information transfer.

1. DNA template strand nucleotide triplets

TAC - CAT - GAG - TAG - ACT

2. Complementary messenger RNA molecule codons

AUG - GUA - CUC - AUC - UGA

3. Amino acids coded for in the growing protein chain

Met - Val - Leu - Ile - Stop

The "stop" at the end of the amino acid chain is coded for by a specific **stop codon**. This codon appears at the end of all series of codons in order to terminate protein synthesis, i.e., to stop the addition of any more amino acids to that particular chain. The codon "AUG" at the beginning of the messenger RNA molecule is the **universal start codon** that places a slightly modified version of the amino acid methionine at the beginning of every newly-formed protein chain in all living things. (However, it is often cut off in a process called post-translational modification that takes place within the cell. This explains why most mature, functional protein chains do not start with methionine).

George the Cat and the Genetic Code

I'm now going to tell you a little story that will help you instantly remember the four letters of the genetic code as well as the hydrogen bonding relationship between the complementary base pairs.

I once had a brown tabby cat named George who was always getting into late-night fights with raccoons in a local park. He was a tough scrapper and had the scars to prove it, but I think he must have met more than his match on one particular occasion because he never came home after I let him out on one fateful night. For this reason, I am going to dedicate this memory trick to him so that his name will live forever in the minds of all students who use this acronym during their studies. (George would have liked it that way!)

One day, my girlfriend Viviane decided to knit George a sweater because she felt that he was cold when he went outside in the freezing winter nights. She decided it would look cute with his name embroidered across the chest part of it, but try as she might, she couldn't squeeze all the letters of his name across the narrow width of the chest of the sweater. So she opted out for the first initial of his name and a full description of our beloved pet: **G.CAT** (for **G**eorge the **CAT**).

One Saturday morning at breakfast, George jumped up onto the kitchen table with his sweater on while I was reading an article on DNA and the genetic code. As I idly stared at him, I suddenly realized that the name on his sweater told a neat little story. G.CAT became G, C, A, T, the four letters of the genetic code. Not only that, but the sequence of the letters GCAT held the key to the complementary base-pairing rule of the nitrogenous bases within DNA, namely that G always pairs with C, and that A always pairs with T!

George the CAT or G.CAT

$$G \equiv C \quad A = T$$
$$ 3 2$$

If you can remember the above visual, **GeeCAT 3 and 2**, you will automatically remember that in DNA, the four bases are Guanine, Cytosine, Adenine, and Thymine. You will also know that Guanine (G) always pairs with Cytosine (C) with three hydrogen bonds, and that Adenine (A) always pairs with Thymine (T) with two hydrogen bonds. This is an important fact to remember when discussing the stronger bond that occurs between G and C because of the extra hydrogen bond. This stronger bonding between the Gs and Cs in the DNA molecule in what are called **GC-rich** regions is what contributes to the regulation of expression of some genes which are not transcribed regularly. The RNA polymerase molecule cannot easily pry apart the more strongly bonded GC-rich regions to begin transcription of that particular gene.

How Do Cells Make Proteins?

As I have mentioned, the sequence of nucleotides in DNA determines the sequence in mRNA which, in turn, dictates the sequence of amino acids in proteins. The first step is called transcription because DNA and RNA are both in the "language" of nucleic acids. The term "translation" best describes protein synthesis because translation means "changing from one language to another." The cell is changing the language of nucleic acids to the language of proteins. You can think of the amino acids as the "alphabet" of the protein language. Since there are 20 "letters" in the amino acid alphabet but only four "letters" in the alphabet of nucleic acids, it takes more than one base to code for one amino acid. Using a tree diagram like the one on page 26, you can see that combinations of two bases give 16 different sequences and that combinations of three bases make 64 sequences. Therefore, it takes at least three bases to create enough sequences to code for 20 amino acids. Three base codes are just what scientists found when they studied the details of translation. Therefore, you could say that the "words" in the language of nucleic acids are three "letters" long.

The three-letter words along the mRNA are called "codons." Each tRNA molecule has a set of three bases, called an anticodon, that "reads" a codon by base pairing. For example, the anticodon CGA on a tRNA will base pair with the codon GCU on a mRNA. The tRNA with the anticodon CGA carries the amino acid arginine. A complete "dictionary," called The Genetic Code, is shown on page 29. You can use this dictionary to find the amino acid attached to the tRNA that base pairs with each of the 64 codons.

Ribosomes, mRNA, and tRNA, as well as many other protein factors, cooperate to translate the information in the mRNA into proteins. Ribosomes are made

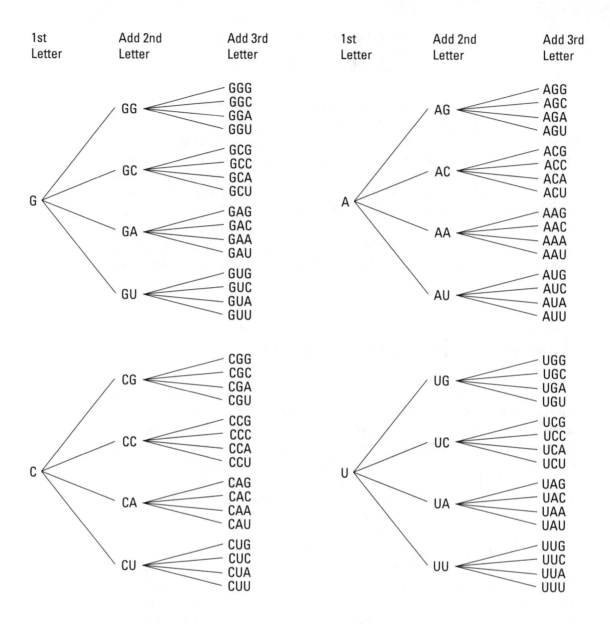

1st Letter	Add 2nd Letter	Add 3rd Letter

This tree diagram shows how combinations of two bases give 16 different sequences, and combinations of three bases make 64 sequences.

in two pieces, a small subunit and a large subunit. The subunits remain separate until translation begins. The four diagrams opposite show how translation occurs to synthesize proteins.

This process continues until one of the three "stop" codons, UAA, UAG, or UGA, reaches the A site on the ribosome. There are no tRNAs that read these codons. Instead, proteins, called "release factors," recognize the codons as STOP codes and cause the entire system to come apart, releasing the new polypeptide chain. The same piece of mRNA can be read over and over to make many identical polypeptide chains.

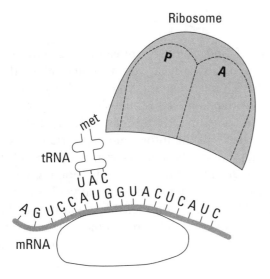

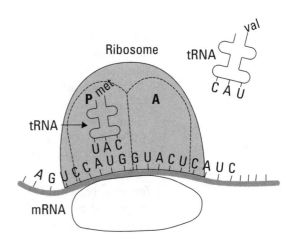

First, several protein initiation factors bind to the small subunit and help it bind to a mRNA in the location of the "start" codon, AUG. The stretch of RNA ahead of the start codon promotes the stable binding of the mRNA to the small ribosomal subunit. The tRNA with the anticodon UAC then hydrogen bonds with the AUG.

Next the large ribosome subunit joins the complex so that the tRNA attaches to a binding site called the "P" site on the large subunit, as shown here.

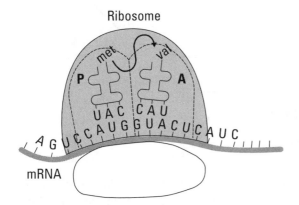

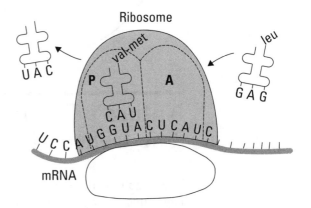

Then a second tRNA with its amino acid binds to the "A" site beside the "P" site. While held in this position, the amino acid on the first tRNA is transferred to the amino acid on the second tRNA as indicated by the arrow.

Finally, the first tRNA leaves the ribosome and the second tRNA moves over into the vacated P site, taking the mRNA with it. The system is now ready for another tRNA to base pair with the next codon and bind to the empty A site.

More Memory Tips

I've given you some acronyms that I find useful for remembering information about DNA. If you'd like to check yourself on what you remember about tRNA and mRNA, draw a Venn Diagram. A Venn Diagram, you say? What is a Venn Diagram? Quite simply, draw two overlapping circles on a sheet of paper. In one circle place all the information about something (in this case, tRNA) that is specific only to it. In the other circle, place all the information specific to mRNA. Then, where the two circles overlap, write the characteristics that tRNA and mRNA have in common. Your Venn Diagram is now complete. Forcing yourself to organize information in this way can help you remember what you've just read or heard. You could also do a Venn Diagram to consider the similarities and differences between DNA and RNA, to help you think through the cellular processes you've been reading about in my little book. I hope you find this so-called graphic organizer, the Venn Diagram, as useful as I have.

Venn diagram

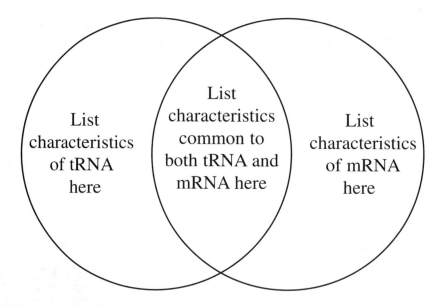

Nature's Recipe for Life: The Genetic Code

The genetic code is made up of 64 DNA nucleotide triplets, shown on the next page, that ultimately code for the step-by-step placement of amino acids into a growing protein chain (as well as coding for the starting and stopping of protein synthesis). On the Genetic Code chart, you will find a complete genetic code dictionary that shows you the following, at a glance: the nucleotide triplets of the DNA coding strand, the DNA template strand, and the complementary messenger RNA strand, plus the amino acid that is ultimately coded for during protein synthesis.

Nature's Recipe for Life

THE GENETIC CODE

complementary messenger RNA Codons

DNA "Template Strand" N-Triplets

DNA "Coding Strand" N-Triplets

Full Name of Amino Acids coded for by the Three-Letter Codons

3 Letter Abbreviations of Amino Acids

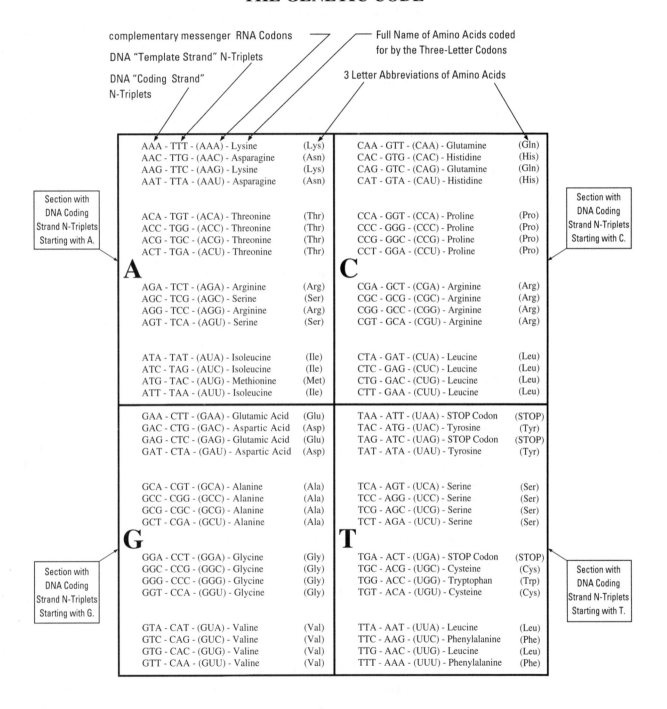

Section with DNA Coding Strand N-Triplets Starting with A.

AAA - TTT - (AAA) - Lysine	(Lys)
AAC - TTG - (AAC) - Asparagine	(Asn)
AAG - TTC - (AAG) - Lysine	(Lys)
AAT - TTA - (AAU) - Asparagine	(Asn)
ACA - TGT - (ACA) - Threonine	(Thr)
ACC - TGG - (ACC) - Threonine	(Thr)
ACG - TGC - (ACG) - Threonine	(Thr)
ACT - TGA - (ACU) - Threonine	(Thr)

A

AGA - TCT - (AGA) - Arginine	(Arg)
AGC - TCG - (AGC) - Serine	(Ser)
AGG - TCC - (AGG) - Arginine	(Arg)
AGT - TCA - (AGU) - Serine	(Ser)
ATA - TAT - (AUA) - Isoleucine	(Ile)
ATC - TAG - (AUC) - Isoleucine	(Ile)
ATG - TAC - (AUG) - Methionine	(Met)
ATT - TAA - (AUU) - Isoleucine	(Ile)

Section with DNA Coding Strand N-Triplets Starting with C.

CAA - GTT - (CAA) - Glutamine	(Gln)
CAC - GTG - (CAC) - Histidine	(His)
CAG - GTC - (CAG) - Glutamine	(Gln)
CAT - GTA - (CAU) - Histidine	(His)
CCA - GGT - (CCA) - Proline	(Pro)
CCC - GGG - (CCC) - Proline	(Pro)
CCG - GGC - (CCG) - Proline	(Pro)
CCT - GGA - (CCU) - Proline	(Pro)

C

CGA - GCT - (CGA) - Arginine	(Arg)
CGC - GCG - (CGC) - Arginine	(Arg)
CGG - GCC - (CGG) - Arginine	(Arg)
CGT - GCA - (CGU) - Arginine	(Arg)
CTA - GAT - (CUA) - Leucine	(Leu)
CTC - GAG - (CUC) - Leucine	(Leu)
CTG - GAC - (CUG) - Leucine	(Leu)
CTT - GAA - (CUU) - Leucine	(Leu)

Section with DNA Coding Strand N-Triplets Starting with G.

GAA - CTT - (GAA) - Glutamic Acid	(Glu)
GAC - CTG - (GAC) - Aspartic Acid	(Asp)
GAG - CTC - (GAG) - Glutamic Acid	(Glu)
GAT - CTA - (GAU) - Aspartic Acid	(Asp)
GCA - CGT - (GCA) - Alanine	(Ala)
GCC - CGG - (GCC) - Alanine	(Ala)
GCG - CGC - (GCG) - Alanine	(Ala)
GCT - CGA - (GCU) - Alanine	(Ala)

G

GGA - CCT - (GGA) - Glycine	(Gly)
GGC - CCG - (GGC) - Glycine	(Gly)
GGG - CCC - (GGG) - Glycine	(Gly)
GGT - CCA - (GGU) - Glycine	(Gly)
GTA - CAT - (GUA) - Valine	(Val)
GTC - CAG - (GUC) - Valine	(Val)
GTG - CAC - (GUG) - Valine	(Val)
GTT - CAA - (GUU) - Valine	(Val)

Section with DNA Coding Strand N-Triplets Starting with T.

TAA - ATT - (UAA) - STOP Codon	(STOP)
TAC - ATG - (UAC) - Tyrosine	(Tyr)
TAG - ATC - (UAG) - STOP Codon	(STOP)
TAT - ATA - (UAU) - Tyrosine	(Tyr)
TCA - AGT - (UCA) - Serine	(Ser)
TCC - AGG - (UCC) - Serine	(Ser)
TCG - AGC - (UCG) - Serine	(Ser)
TCT - AGA - (UCU) - Serine	(Ser)

T

TGA - ACT - (UGA) - STOP Codon	(STOP)
TGC - ACG - (UGC) - Cysteine	(Cys)
TGG - ACC - (UGG) - Tryptophan	(Trp)
TGT - ACA - (UGU) - Cysteine	(Cys)
TTA - AAT - (UUA) - Leucine	(Leu)
TTC - AAG - (UUC) - Phenylalanine	(Phe)
TTG - AAC - (UUG) - Leucine	(Leu)
TTT - AAA - (UUU) - Phenylalanine	(Phe)

In any textbook that I've ever looked at, the only type of genetic code dictionary that I've seen is presented in a format where you match up the letters of the messenger RNA molecule at the sides of the chart to locate the amino acid that is coded for somewhere in the middle. So I've made it a lot easier for you to zero in on the amino acid that is coded for by giving you a visual of each of the 64 nucleotide triplets of the two DNA strands as well as the codons of the messenger RNA and the amino acids for which they code.

In the past there has been a bit of confusion surrounding the nomenclature of the DNA coding strand and the DNA template strand. As you know by now, the DNA molecule is made up of two individual strands that wrap around each other to form the double helix. When the strands unwind for the process of transcription to begin, only one of the two strands in any given region of DNA is transcribed onto a messenger RNA molecule. The stretch of nucleotides at that same region on the other strand is not transcribed onto mRNA. The DNA strand that goes through the transcription process is called the DNA template strand because it serves as the mold or template for the formation of a complementary strand of messenger RNA. The other DNA strand is called the coding strand because its nucleotide sequence is the "same" as that of the messenger RNA molecule that carries the code to be directly translated into a protein chain on the ribosome. (The only exception to this rule is that wherever a Thymine (T) appears in the DNA coding strand, there is a Uracil (U) in the RNA strand).

To see this for yourself, take a look at The Genetic Code chart. You will notice that the nucleotide triplets in the first column, labeled as the DNA coding strand nucleotide triplets, are the same as the nucleotides in the third column labeled as the complementary messenger RNA codons, except that a T in the DNA is a U in the messenger RNA.

Please note that in current textbooks and scientific publications, whenever the nucleotide sequence of a gene is written in terms of DNA nucleotide triplets, it is the nucleotide triplets of the DNA coding strand (also referred to as the +strand or sense strand) that are used, not the nucleotide triplets of the DNA template strand. Please also note that on the Genetic Code chart, the term nucleotide triplets has been abbreviated to N-triplets.

Some Important Events in the History of Genetics

1859   Charles Darwin publishes his theory of evolution in the book, *On the Origin of Species*.

1865   In an obscure science journal, Austrian monk Gregor Mendel publishes his results of breeding experiments with the garden pea plant. He discovered that physical characteristics are transmitted from parents to offspring as discreet unit factors.

1884   Gregor Mendel dies an unsung hero because his discovery of the mechanism of heredity is largely ignored by the scientific community.

1900 Mendel's work is rediscovered and confirmed by experiments done by Dutch botanist Hugo deVries and German botanists Karl Corens and Erik von Tshermak-Seysenegg.

1905 English botanist William Bateson invents the term **genetics** to describe the study of inherited traits.

1909 Danish plant physiologist Wilhelm Johannsen introduces the term **gene** to refer to the discrete units of heredity.

1910 American scientist Thomas Hunt Morgan (winner of the 1933 Nobel Prize) discovers that the occurrence of white eyes in fruit flies is linked to the sex of the fly. This correlation paved the way for the discovery of more **sex-linked** inheritance patterns.

1911 Thomas Hunt Morgan's breeding experiments propose that white eyes, yellow body, and miniature wings in the common fruit fly are indeed linked together on the X-chromosome (the female sex chromosome).

1928 British scientist Frederick Griffith discovers the phenomenon of **transformation** (i.e., the transfer of genes between cells) in a strain of pneumonia-causing bacteria called *Diplococcus pneumoniae*.

1941 American scientists George Beadle and Edward Tatum (winners of the 1958 Nobel Prize) establish the **one gene – one enzyme** concept with work done on the red bread mold, *Neurospora crassa*.

1944 American scientists Oswald Avery, Colin MacLeod, and MacLyn McCarty prove that the transforming principle (discovered by Griffiths) of pneumococcal cells is DNA and not protein.

1950 American scientist Barbara McClintock (winner of the 1983 Nobel Prize) postulates the theory of **jumping genes** that determine the color of corn kernels in multicolored Indian corn.

1952 American scientists Alfred Hershey (winner of the 1969 Nobel Prize) and Martha Chase demonstrate that the genetic material of a bacteriophage (a virus that attacks bacteria) is indeed DNA and not protein.

1953 American scientist James D. Watson and British scientist Francis H. Crick (winners of the 1962 Nobel Prize) use the X-ray diffraction data of Rosalind Franklin and Maurice Wilkins (co-winner of the 1962 Nobel Prize) as well as the base-pairing rule data of Erwin Chargaff to deduce the 3-dimensional structure of the DNA double helix.

1956 American scientist Arthur Kornberg (winner of the 1959 Nobel Prize) isolates DNA polymerase I from *Escherichia coli* bacterial cells.

1958 American scientists Matthew Meselson and Franklin Stahl demonstrate that DNA replication occurs in a semiconservative fashion (see Glossary for definition of semiconservative).

1959 Severo Ochoa (winner of the 1959 Nobel Prize) discovers the first RNA polymerase.

1961 French scientists Francois Jacob and Jacques Monod (winners of the 1965 Nobel Prize) propose the lactose operon as a model to explain the regulation of gene expression in *E. coli* cells during the metabolism of the milk-sugar called lactose.

1961   Francis H. Crick, Sydney Brenner, and their colleagues propose that the genetic code is made up of three consecutive nucleotides that lie side-by-side along one strand of DNA.

1965   American scientist Robert Holley (winner of the 1968 Nobel Prize) is the first to determine the complete sequence of a transfer RNA molecule (the transfer RNA molecule specific for the amino acid alanine).

1967   The work of American scientists Marshall Nirenberg and Har Gobind Khorana (winners of the 1968 Nobel Prize) results in the complete cracking of the genetic code.

1970   American scientists Daniel Nathans and Hamilton Smith (winners of the 1978 Nobel Prize) discover the first restriction enzymes in bacterial cells.

1970   American scientists David Baltimore and Howard Temin (winners of the 1975 Nobel Prize) discover the enzyme reverse transcriptase in RNA-containing viruses (retroviruses) that cause tumors in animals.

1972   American scientist Paul Berg (winner of the 1980 Nobel Prize) performs the very first experiment that spliced together DNA from two different sources (the creation of the very first **recombinant DNA** molecule).

1973   American scientists Herbert Boyer and Stanley Cohen (winner of the 1986 Nobel Prize) transplant a gene from a toad into a bacterial cell and get expression of the toad gene within the bacteria.

1977   American scientists Allan Maxam and Walter Gilbert (winner of the 1980 Nobel Prize) and British scientist Frederick Sanger (co-winner of the 1980 Nobel Prize) publish their techniques for sequencing the nucleotides within DNA.

1980   The United States Patent Office grants the General Electric Company the world's first patent on a genetically engineered bacterial strain that "eats up" oil spills. The bacterium was developed by Dr. Ananda Chakrabarty.

1982   The world's first genetically engineered drug is made available by the Eli Lilly International Corporation. It is human insulin, produced by genetically engineered bacterial cells and sold under the name "Humulin."

1984   American scientist Philip Leder produces the world's first genetically engineered animal. He implanted human cancer-causing genes into the chromosomes of a mouse.

1988   American scientist James D. Watson (winner of the 1962 Nobel Prize) takes on the job as chief of the Human Genome Project.

1989   Francis Collins, John Riordan, and Lap-Chee Tsui discover and isolate the gene that causes cystic fibrosis.

1990   American researchers W. French Anderson and R. Michael Blaese successfully conduct the world's first gene therapy experiment at the National Institute of Health in Bethesda, Maryland, on a little girl suffering from an immune system disorder.

1995   Peter St. George Hyslop and his colleagues at the University of Toronto discover and isolate the gene responsible for early-onset Alzheimer's disease, located on chromosome #14.

A Little Bit of DNA History

DNA	Biotechnology	Genetic Engineering
3.5 billion yrs. old	At least 10 000 yrs. old	About 25 yrs. old

The approximate ages of DNA, Biotechnology, and Genetic Engineering.

DNA is believed to have been on this planet for at least 3.5 billion years. Fossils were found in Western Australia of very primitive early cyanobacteria that are presumed to have contained DNA as all present-day cells do. These Precambrian-era fossils have been dated at about 3.5 billion years old. This means that the DNA molecule has kept its secret of the genetic code locked away for a very long period of time!

Biotechnology

The term "biotechnology" was coined in the late 1800s and has come into general use relatively recently. However, the practice itself has been around for at least 10 000 years. Around this time, primitive farmers performed the first applications of genetics by selectively breeding plants and animals. About 10 000 years ago, humans evolved culturally from being solely hunters and gatherers of food to growers of plants and domesticators of animals for the purpose of controlling their food supply. It probably happened by chance at first, when food gatherers discovered cereal plants such as corn, wheat, barley, and oats growing wild in areas that were spread out over many square miles of land. After many generations of using these plants for nourishment, early people began to keep seeds from the plants, planting them in a concentrated fashion, covering a smaller land mass. Thus were established the first farm fields. Groups of wandering, nomadic people were able to settle down beside their farm fields and develop villages and therefore a more sedentary lifestyle.

The earliest farmers realized that physical traits of these food plants, such as size, color, shape, and taste, were somehow carried on to the next generation by the seeds. They took special notice of these traits, plus nutrient value and resistance to conditions such as drought and insect infestation, and planted the seeds of plants that exhibited more desirable characteristics. This process of planting seeds from the best plants, and mating animals with the best traits (at least those perceived as "best") is called **selective breeding**.

Over several thousand years of selective breeding, crop plants became bigger and tastier and gave higher yields. Certain cereal crops were responsible for the development of civilizations in certain regions of the world. For example, the Middle East and southern Europe benefitted from the development of better strains of wheat. Similarly, the growth of Asia was enhanced by better strains of rice while improved strains of corn helped to develop the nations of the Americas. By about 5000 years ago, just about every food crop that is cultivated in the world today was being selectively bred in some part of the world.

Most people think that biotechnology means manipulating or controlling a living cell for the production of some organic substance that is useful to mankind. This definition is true (along with the technology of selective breeding). An interesting fact is that the very first application of this type of biotechnology dates back to about 8000 years ago. That's the age of the world's first recipe for brewing beer! Archeologists have discovered hieroglyphics depicting ancient Babylonians fermenting barley in a pot and storing the brew. These hieroglyphics, as well as other historical artifacts, strongly suggest that this was the first use of live yeast cells to ferment sugar into alcohol.

Genetic Engineering

Genetic engineering is the manipulation of DNA and RNA of living organisms. Simply put, genetic engineers take genes from one species and put them into another, so that the same gene will be expressed in a different organism. The science of genetic engineering is relatively new — about 25 years old. In 1970, Dr. Hamilton Smith and his colleagues at Johns Hopkins University in Baltimore, Maryland, discovered a unique group of enzymes that cut DNA molecules in very predictable and reliable ways. The enzymes (called **restriction enzymes**) were discovered in certain types of bacterial cells; they cut-up or "restrict" any foreign DNA molecules that somehow manage to find their way into the cell. (Usually, the foreign DNA would be from a bacteriophage attempting to infect the cell.) It was further shown that these restriction enzymes would cut DNA from any source at exactly the same sequence of nucleotides (called "restriction sites" or "recognition sequences"). The cut DNA usually had what are called **sticky ends**: single-stranded ends of nucleotides that were complementary to each other. Any population of DNA molecules cut with a specific type of restriction enzyme that left sticky ends could be joined to each other. A very important point to highlight here is that if you used restriction enzyme "A" to cut DNA from one species of organism (a bacterial cell, for example), then used the same restriction enzyme "A" to cut DNA from another species of organism (a toad, for example), the freshly cut pieces of DNA would have complementary sticky ends. Therefore, you could join the DNA from the toad to the DNA from the bacterial cell. In so doing, you would be splicing together gene-carrying pieces of DNA that in nature would never be connected with each other in such a close structural relationship because of the barrier of hundreds of millions of years of evolution separating the bacterial cell from the toad.

The example used above, of DNA from a toad being spliced to the DNA from a bacterial cell, is the actual experiment that was performed in 1972 by two groups of scientists in California. The amazing part of this construction of a **recombinant DNA** molecule is that although the toad DNA carried genes for the production of toad proteins, the toad gene was found to function normally inside the bacterial cell! The scientists who led this remarkable research project are Dr. Herbert Boyer and Dr. Stanley Cohen. Dr. Cohen was awarded the 1986 Nobel Prize for Medicine or Physiology.

Now, it's not by accident that I put DNA, biotechnology, and genetic engineering together in this final discussion because DNA is at the heart of all biotechnology and genetic engineering.

The Heart of All Biotechnology is the DNA Molecule

Biotechnology is the use of living systems such as single cells to produce desirable proteins and other organic molecules that are useful to humankind. DNA is at the heart of all biotechnology because DNA is the blueprint for all protein molecules!

Look again at the list of the eight major types of proteins (on page 7). With this list as a guide, you will be able to recognize immediately the significance of most new developments in biotechnology by zeroing in on the category of protein that the research effort is trying to produce.

The AIM of Biotechnology

Three main areas will be most greatly impacted by biotechnology: Agriculture, Industry, and Medicine. I like to use the acronym **AIM** to emphasize this point in the following way:

The AIM of biotechnology is to enhance the quality and quantity of proteins that are of practical value in the areas of:

A   I   M

Agriculture   Industry   Medicine

Agriculture

Let's first look at how biotechnology and genetic engineering are applied to the field of agriculture. The efforts of DNA manipulation in this field can be broken down into two areas, plants and animals.

Plants

One major thrust in plant genetic engineering is focused on the so-called "grass-plants" which we know as the major food crops such as barley, corn, wheat, rice, and oats.

The higher plants are divided into two major classes: the monocotyledons and the dicotyledons. When the germinating seed bursts open, the newly developing plant will possess either one or two tiny seed leaves. If it has only one, it is classified as a monocotyledon. If it has two tiny seed leaves, then it is a dicotyledon.

Grass-plants happen to be monocotyledonous plants, and modern genetic engineering efforts are focusing much attention on the development of more reliable and effective ways to introduce foreign genes into monocotyledons. To be able to manipulate the genes of these monocots is an important science to master, since these plants (barley, corn, wheat, oats, and rice) compose the basic diet of much of the world's population. Genetic engineering is focused on making these crop plants produce higher yields with increased nutrient value.

Genetic engineering technology is also attempting to do the following: make crop plants resistant not only to insect pests but also to viral infections for which they have no natural immunity, help plants grow under harsh environmental conditions such as prolonged droughts or poor soils, and decrease fertilizer use by genetically altering plants so they take nitrogen from the atmosphere and "fix" it into nitrates that the roots of the plant can absorb. (The nitrogen can then be incorporated into amino acids and nucleotides that are the building blocks of the plant's own proteins and nucleic acids.) If all of these goals are met by biotechnology, it would result in a tremendous increase in the food supply for an ever-increasing world population.

Perhaps the most important advance in plant genetic engineering that has been accomplished to date is delaying the over-ripening (or spoiling) process of tomatoes. These tomatoes ripen normally, but remain ripe without spoiling for seven to ten days longer than ordinary tomatoes. Therefore, they can be allowed to ripen on the vine and shipped to market without spoiling. If this technique can be successfully applied to all food crops, then the food can travel from farm fields to hungry mouths before it rots (a very significant point since at present about half of the world's food rots or is consumed by insects before it reaches hungry people).

Animals

We can draw examples of genetic engineering applications from at least four types of farm animals: cattle, pigs, sheep, and chickens.

Cattle

Two types of cattle important for food production are dairy cattle and beef cattle. Genetically engineered bovine growth hormone, called Bovine Somatotropic Hormone (BST), is injected into dairy cattle in some countries for the purpose of increasing milk production by as much as 20%. BST is also injected into beef cattle (also only in some countries) to produce higher yields of leaner meat.

Pigs

Human growth hormone (and porcine growth hormone) has been genetically engineered into pigs to make them grow faster and produce leaner meat.

Sheep

Some very sophisticated genetic engineering techniques have been applied to produce what are called **transgenic sheep**. This involves injecting human genes into the fertilized egg of the sheep so that each and every cell derived from that original zygote contains the foreign gene. This means that in every cell of the adult sheep, including the sex cells, the artificially inserted human DNA is present. The reason for this effort is to have the human gene for a blood-clotting protein, called Factor IX, become expressed in the cells of the milk-producing glands of the sheep. Thus, the human protein is secreted into the milk and can be isolated and purified. Factor IX is a very important protein in the blood-clotting process and is deficient in some hemophiliacs.

This technique of producing transgenic animals could be applied using any number of desirable human proteins that are naturally secreted in only very small amounts by humans. By producing transgenic farm animals that secrete these human proteins into their milk, the aim is to produce much larger amounts of these scarce and expensive human proteins for use in medical treatment of a wide range of diseases.

Chickens

In 1989, the world's first healthy transgenic chicken was produced by a California-based genetic engineering company. The inserted genes are also passed on from one generation to the next. Foreign genes were added to freshly laid eggs by first inserting the genes into a virus particle that normally infects chicken eggs. The virus particle with the foreign gene was rendered harmless and allowed to infect the eggs. It injected its recombinant DNA (its own now harmless DNA plus the newly added DNA) into the eggs.

Scientists hope that three kinds of transgenic chickens would be produced by these experiments. The first type would be resistant to certain viruses that spread quickly and wipe out a significant percentage of chickens in broiler houses. The second type would have an added growth hormone gene in their cells to make them grow faster. The third type of transgenic chicken would produce valuable medical proteins in their eggs that could be used for treatment of human disease. How could this be done? If you think about it, what do you see when you crack open an egg onto a hot frying pan? You see the coagulation of a lot of egg white protein called albumen. In fact, albumen makes up a large percentage of the total amount of protein in a freshly laid egg. Because albumen is made in very large amounts, it means that the gene for albumen is very strongly expressed. This strong expression is regulated by a genetic element called a **promoter**. If you could somehow add to the egg a gene that codes for a valuable human protein, so that the new gene is under the control of the strong albumen promoter, then you could get the hen to produce large amounts of this desirable protein in the albumen. This would greatly increase the availability of some very scarce human proteins such as interleukins, which are naturally secreted by white blood cells in order to boost the entire immune system.

Industry

The applications of biotechnology in industry are too numerous to mention in a brief synopsis such as this one, so I'll pick just a few to give you an idea of where it might lead. One of the most well-known uses of living organisms in industry is the use of naturally-occurring bacterial cells that "eat up" oil spills. Researchers are attempting to construct more efficient oil-eating bacteria by genetic engineering.

Soil bacteria present a rich source of potentially beneficial workers in industry. As you know, bacterial cells multiply at a rapid rate, which can result in the accumulation of many mutations in the bacteria population over a short period of time. If bacteria are collected from soil which has been polluted with toxic waste materials over many years, it is probable that a strain of these bacteria might be able to feed on the toxic chemicals. Bacteria can potentially use any carbon-containing compound, even a toxic pollutant such as polychlorinated biphenyls (PCBs), as food, breaking them down to non-toxic by-products. Genetic engineers are isolating these pollution-eating bacteria, and through genetic transfers between different mutant strains, strengthening their appetites for any given toxic material. This would greatly enhance the clean-up of industrial sites polluted with chemical waste materials.

Another major thrust of biotechnology application in industry is the relatively new development of what are called **antibody enzymes** or **abzymes**. Many industrial processes require metallic catalysts such as platinum or palladium to speed up the rate of chemical reactions. These reactions often require very high temperatures and pressures. Enzymes are biological catalysts which speed up the rate of biochemical reactions in a living cell at the temperature and pressure of the organism. It stands to reason that if an industrial reaction could be catalyzed at around 37°C and 100kPa pressure (98°F and 1 atm.), instead of at 2000°C and 500 kPa (3600°F and 5 atm.), then a great deal of energy could be saved in completing that particular chemical reaction. Saving energy is a constant concern in industry because energy costs money! For this reason, any branch of industry that performs catalyst-mediated chemical reactions is looking very closely at the use of enzymes instead of conventional catalysts.

Now that we have established the practicality of enzymes in industrial chemistry, let's get back to how the abzymes will play a vital role. Like enzymes, antibodies are very large protein molecules produced within higher life forms such as mammals. Like enzymes, antibodies are built with a specific shape that allows them to interact with another molecule of a specific complementary shape. However, unlike an enzyme, an antibody's function in an organism is not to speed up a chemical reaction. Its job is to hunt down a molecule that could be a part of a germ that has entered the bloodstream of the animal and inactivate that germ. What genetic engineering is slowly making possible is the production of antibody molecules with unique 3-dimensional shapes that will allow them to act as biological catalysts in industry. What makes genetically engineered antibodies so interesting to biotechnology is that a living mammal such as a mouse or a human can make well over 100 million different types of antibodies, each one having its own special kind of binding specificity. This vast structural diversity in abzymes will phenomenally increase the repertoire of catalysts that will be available for the industry of the future.

A Biological Electronic Microchip?

The final industrial example that I will briefly detail is one that I think is the most intriguing and far reaching: the biological electronic microchip. The fastest computers in the world today have their incredible information-processing speeds thanks to the microchips that serve as the "brains" of the computer. The more electrical signals that the chip can take in at one end, the faster is the processing speed of the computer. One of the limiting factors in a microchip's ability to process information is that the electrical input generates heat within it. If you try to process too much information in the form of electrical signals through the microchip, it will absorb too much heat and melt down. This is especially true as the microchips of today are being made smaller and smaller through miniaturization. As more electrical signals are passed through a smaller, more condensed chip, the hotter it gets. That is why today's desktop computers with the ultrafast microprocessing chips come equipped with larger fans built into the central processing unit to cool them down.

Some laboratories around the world are studying the processing of electrical signals through protein molecules. Some proteins have been shown to absorb light energy and convert it into an electrical current. This field is of tremendous interest to people who build microelectronic circuitry as well as optical circuitry, in which the signals are transmitted by beams of light instead of electrons. At this stage of the game, proteins certainly don't do the job of processing light beams and electrical currents in a way that is of practical use. However, the stage is being set for altering genes to code for proteins that will be able to handle all of the electro-optical signals required of computers of the future. For example, genetic engineering could produce proteins that will rapidly pass signals that originate from either a light source or an electrical current. These protein molecules could then be designed to attach themselves to each other and become aligned into orderly rows on a solid surface resembling a parallel processor. If proteins could be used as microchips they would be thousands of times smaller than the chips of today. Unlike present-day microchips that absorb heat and require cooling, proteins retain very little if any heat at all when transmitting an electrical current or light energy. This means that they would work much faster and could be much smaller.

A branch of science that is rapidly commanding attention by industry is called **nanotechnology**. This field focuses on miniaturization of mechanical and electrical parts to the atomic or at least the molecular level, with the goal of cramming more power into less space. In fact it is easy to hypothesize that once these protein-microchips become a reality, the computing power of a football-field covered with present-day supercomputers will be condensed into something the size of a sugar-cube. This blending of genetic engineering, electro-optics, and semi-conductor science is the stuff of which technological revolutions are made!

Medicine and Forensic Science

The main areas in which biotechnology will make advances in the field of medicine are the Human Genome Project, the diagnosis of genetically-based diseases, forensic science, the production of proteins of pharmaceutical significance, and gene therapy.

The Human Genome Project

This project is an international effort, headed by the USA. It was officially initiated in 1990 and its projected completion date is the year 2005. Its goal is to map the precise location of all of the estimated 100 000 or so genes that reside within the DNA of the human genome. If this task is accomplished, then scientists will be in an excellent position to identify every gene that causes human disease and perhaps cure specific conditions by gene therapy.

The human genome is comprised of one complete set of 23 chromosomes. This may seem confusing to you since you probably know that a normal human being has 46 chromosomes in each cell. The number 46 represents two homologous sets of 23 chromosomes each. Therefore one set is made up of 23 chromosomes. (See "homologous" in the Glossary.)

There are an estimated three billion nitrogenous base pairs in the human genome (i.e., in one set of 23 chromosomes). As you can imagine, it will take a tremendous amount of time and effort to sequence all three billion base pairs. However, once the precise location of each gene has been determined with a complete sequence map, then any human gene can be isolated and cloned for the sake of close study and greater understanding of its function in the human body.

Diagnosis of Genetically-Based Diseases

There are more than 3000 known human diseases that can be genetically inherited. Scientists are now able to diagnose more than 200 genetic disorders through DNA technology. These diagnoses can be made even before a baby is born, by performing a genetic screening test on the fetus's cells, collected from the pregnant mother's womb by amniocentesis or chorionic villus sampling (see Glossary). Prenatal diagnosis is very important because early treatment to a newborn child, or even to a fetus, could prevent suffering or death.

One example of this kind of early diagnosis/preventive treatment is in the case of the recessive genetic disease called phenylketonuria. This disease is caused by a single-gene defect. The gene in question codes for an enzyme called phenylalanine hydroxylase, which converts the amino acid phenylalanine to the amino acid tyrosine. If this enzyme is defective, a build-up of phenylpyruvic acid causes brain damage and a smaller-than-normal head-size. All infants in the US and Canada (and in many other countries) are routinely tested for this defect shortly after birth. If this single-gene defect is detected, the newborn child can be fed a diet that is low in phenylalanine, preventing mental retardation and promoting normal development.

Forensic Science

DNA technology is taking on a major role in criminology. It is dramatically adding to the repertoire of police investigators in identifying guilty individuals who wear gloves and leave no fingerprints at the scene of a crime. As long as the guilty party leaves just one hair with an intact root tip, a small drop of blood, or a few skin cells at the scene of the crime, DNA can be extracted from these samples.

The DNA in the sample is then amplified by using the polymerase chain reaction (PCR) (see Glossary). The PCR technique serves as a "photocopier" of DNA. A billion copies of a piece of double-stranded DNA can be made in a matter of a few hours. With a sample of a billion copies of an original piece of double-stranded DNA to work with, certain identification tests can be performed to compare the DNA found at the scene of the crime with the DNA of a criminal suspect. If samples match, then the probability is extremely high that the DNA left at the crime scene belongs to the suspect. This identification process is called "DNA Fingerprinting" (but should not be confused with the print patterns that exist on the tips of one's fingers).

This technique is also effective in paternity cases in which the true father of a child is being determined. Since half of the child's DNA was contributed by the father, half of the child's DNA pattern will match with half the DNA pattern of the true father. A male who is not the biological father of the child will have a DNA pattern that shows very little similarity to the child's pattern.

The Production of Pharmaceutically Important Proteins

Biotechnology is even now increasing the quality and the quantity of medically useful proteins. The first human protein to be made in genetically engineered bacterial cells was the hormone insulin. Some diabetics experienced severe allergic reactions to pig or cow insulin, the insulin that was previously available to control their blood sugar levels. The gene for human insulin was cloned in *E. coli* bacterial cells, in which it was mass produced (see gene cloning in the Glossary). The insulin produced by the bacteria proved to be more compatible for people unable to take insulin extracted from animals.

Another important protein that has been genetically engineered in relatively large amounts is human growth hormone. Prior to gene cloning, human growth hormone was extracted in very small amounts from the pituitary glands of numerous human cadavers. In at least one case, growth hormone removed from a single cadaver infected several people with a dangerous virus. Growth hormone made from genetically engineered bacteria would not pose that problem.

Human interferon, which protects against viral infection, was previously extracted from the blood in such tiny amounts that it was difficult to get enough of it even to study its mode of action and potential effectiveness against a variety of test viruses. Now, much larger quantities are obtained from genetically engineered bacteria.

Factor VIII was previously isolated from human blood samples in a very laborious and expensive process. The fact that the blood from which it was derived could have been contaminated with viruses such as the virus that causes AIDS made it possible that the Factor VIII sample was also contaminated. The cloning of the Factor VIII gene and its subsequent expression in mammalian cells in tissue culture eliminated the possibility of human blood-virus contamination.

Vaccines are usually produced by preparing inactivated virus particles which are then injected into a person. The vaccine stimulated the immune system to ready itself in case of infection with a live version of that inactivated virus. One problem with this method is the possibility that the initial preparation of the inactive virus could contain some still virulent (live) virus particles that could cause serious disease in the vaccinated person. Scientists are now cloning the gene for a single viral coat protein. If it serves to stimulate the immune response, the risk of a live virus contaminating the vaccine preparation is removed.

Gene Therapy

DNA technology has come so far that it has the potential to correct certain types of genetic diseases in humans. This would be accomplished by replacing or supplementing a defective gene with a "good" gene that codes for a well-defined protein that functions normally within the human being. For the foreseeable future, this kind of gene therapy would lend itself only to those diseases that are caused by a single defective gene. (Many genetic disorders are caused by the interaction of several genes, such as diabetes mellitus which results in a blood-sugar imbalance). Some examples of single-gene defects that may be treated by gene therapy are:

1. Dwarfism (caused by a defect in the single gene that codes for the growth hormone protein).

2. Sickle-cell anemia (caused by a defect in the single gene that codes for the beta chain of hemoglobin molecules).

3. Hemophilia (caused by a defect in the single gene that codes for the human blood-clotting-factor protein called Factor VIII).

4. Phenylketonuria (caused by a defect in the single gene that codes for the enzyme protein called phenylalanine hydroxylase).

Scientific history was made in 1990 when a four-year-old girl received the very first gene therapy at the National Institute of Health in Bethesda, Maryland. She suffered from a defective single-gene disease called SCID (severe combined immunodeficiency disease). In this condition, the gene that codes for the enzyme protein called adenosine deaminase is defective. Children with this disease lack the proper form of this enzyme that assists in the breakdown of adenosine. Without the enzyme, toxic by-products of adenosine build up in a group of white blood cells called T-lymphocytes that play an important role in the immune system. This defect eventually destroys the victim's immune system and the child has no defense against infections.

In the procedure that was carried out under strict supervision, the little girl's own T-lymphocytes were removed from her bloodstream. These T-lymphocytes were incubated in a culture dish with a virus that had been rendered harmless and had been genetically engineered to carry the gene for a normally functioning adenosine deaminase enzyme. The virus then injected the "good" gene into the girl's T-lymphocytes. This freshly injected recombinant DNA became spliced into the DNA of the T-lymphocytes. The cells were then returned to the patient's body where they began to produce the normal form of the adenosine deaminase enzyme. The little girl's immune system began to show remarkable improvement after several such treatments. The next step will be to use the viral injection system to inject "good" genes into immature bone marrow cells. This would result in a much broader and longer-lasting treatment since the bone marrow cells give rise to all of the cells of the immune system.

So far, gene therapy has been conducted on human somatic cells only. There is general concern among the scientific community about performing these genetic engineering experiments on germ-line genes (i.e., the genes of the sex cells) because the newly formed gene configurations would be passed along from generation to generation. It is argued by many that the genetic manipulation of germ cells is wrong because it would affect the physical characteristics of future generations of humans. There is concern that this could possibly lead to a revival of the practice of eugenics, which is the deliberate effort to produce human populations with what are believed to be more desirable physical traits. For this reason, a moratorium is in place regarding genetic engineering experimentation on human germ-line cells.

Some Pros and Cons of Genetic Engineering

Concerns

Much has been written concerning the pros and cons of manipulating the genetic code of DNA. Many qualified individuals (among them leading scientists), who are capable of giving educated opinions about DNA technology, are outspoken critics against many applications of genetic engineering. Their views generate a healthy debate and should be taken into consideration as a balance against unilateral decisions to go ahead with any genetic engineering applications by the proponents of the technology. Some opponents of genetic engineering think that a newly formed "super-bug" type of bacteria could escape from the laboratory and spread global epidemics of uncontrollable diseases.

Critics also warn against the release of genetically engineered organisms into the environment because it could result in disastrous disturbances in the balance of ecology. After all, some argue, if you release a genetically altered trout into the streams of the world it will reproduce and pass the new genes on to future generations of trout. If you realize that you've made a mistake because this new "super-fish" alters the delicate balance of the food chain, you can't just recall all the fish that contain the recombinant DNA as you could recall a thousand trucks with a faulty rear axle. Any genetically engineered organisms that have been released into the environment, including the mistakes, would be out there to stay.

Other critics suggest that the future course of genetic engineering could provide a doomsday weapon of biological warfare to both military leaders and terrorists around the world.

Sociologists and philosophers also offer some interesting viewpoints on genetic engineering that touch on their strong disagreement with the "mechanization and thrusting of industrial engineering principles" on life itself. Jeremy Rifkin, one of the world's most outspoken critics of gene manipulation, has said that the term "engineering" implies the design and production of a thing, with the application of standards of quality control. He feels that applying engineering standards of "quality control" to living organisms is morally unacceptable.

All of the above ideas and beliefs are thought-provoking and deserve consideration. The social, legal, ethical, and moral implications of gene manipulation are many and will surely provide a great deal of controversy and debate in the years to come.

Promises

As for the "pro" side of the coin, genetic engineering may offer some of the most versatile and effective solutions to serious problems that face humankind now and in the future. Among these are the many developments in agriculture, industry, and medicine discussed in the previous section. It has been estimated that by the year 2025, the world's population will have swollen to more than 8.5 billion people. With this looming population crunch comes the harsh reality of a diminishing supply of five vital components necessary for human survival: available land space, and sufficient supplies of food, air, fresh water, and energy. (In time, the limited quantity of fossil fuels buried beneath the earth will become exhausted).

As already mentioned in the section on Agricultural applications, the development of higher yielding and more nutritious strains of food plants that ripen more slowly can tremendously increase the available food supply within the next 30 or so years. In terms of increasing the world's fresh-water supply, it is conceivable that microbes can eventually be genetically engineered to desalinate the salt water of the seas and the oceans to produce fresh water fit for human consumption. Many feel that solutions to these and other key problems depend on new genetic engineering applications.

Glossary

In the following pages of terms and definitions, I have tried to compile a comprehensive list that I felt would be useful to anyone studying DNA at the senior high school level, throughout a university biology program, or simply on one's own. It should serve you as a handy reference guide for several years. And, who knows, this may be the only book you ever use that has a Glossary as extensive as its text!

For some of the definitions, I have given a fairly lengthy and detailed explanation to give you a synopsis of a topic that is often discussed in less detail in most textbooks. In order to help you focus on the terms that are most relevant to you, I have put an asterisk beside the entries that are compatible with a university-level biology program. Those entries that are unmarked will be useful to you in high school and as a review in university.

* A-DNA: There are three forms of DNA: A-DNA, B-DNA, and Z-DNA. By far the most commonly occurring form in living cells is B-DNA (i.e., the classical Watson-Crick model of DNA). A-DNA is formed by dehydration of B-DNA. Like B-DNA, A-DNA is also a "right-handed" double helix. If you could look down the long axis of the double helix you would see that the two chains wind around each other in a clockwise (or right-handed) direction, with about 11 nucleotides per 360° turn. The planes of the base pairs form an angle of 20° with the plane perpendicular to the axis of the double helix. (See B-DNA and Z-DNA).

A site: See aminoacyl site.

* abzymes: Monoclonal antibodies that act as catalysts. They are specific for certain substrate molecules. One binds to a substrate molecule, shaping it into the transition state that it has to go through to become a certain product. Much focus is being aimed at abzymes in genetic engineering for use as catalysts in a wide range of industrial processes.

adenine: One of the four nitrogenous bases, found in DNA and RNA.

AIDS: Acquired Immunodeficiency Syndrome. The term given to the disease state of the immune system at the late stage of HIV infection. It is characterized by a very low "helper T cell" count (helper T cells are one of several different types of white blood cells that play an important role in the immune system). A low white blood cell count renders the body susceptible to infection by opportunistic bacteria such as those that cause pneumonia.

albino: An organism that lacks the pigmentation that cause normal coloration. In humans, for example, albinos lack the pigment melanin, a deficiency caused by a genetic mutation. The defective gene results in the lack of the enzyme, called tyrosinase, that plays a key role in the biochemical pathway that ultimately produces melanin. Albinism is a homozygous recessive genetic condition.

allele: One of several alternative forms of a given gene.

* amber codon (UAG): One of the three STOP codons of messenger RNA that code for termination of protein synthesis. The other two mRNA codons that terminate translation are UAA and UGA. UAA is called an "ochre" codon while UGA is called an "opal" codon. The terms "amber," "ochre," and "opal" bear no association with an actual color; they were chosen arbitrarily by the originators.

* amber mutation: A term used in molecular biology referring to a mutation at the DNA template level that results in an amber mRNA codon (UAG). Before the mutation, that particular codon coded for an amino acid. For example, if a nucleotide triplet in a DNA template strand originally read as "GTC," it would code for the mRNA codon CAG which in turn would be translated into the amino acid glutamine. Now, if the DNA nucleotide triplet is mutated to "ATC," then the new mRNA codon will be transcribed UAG (i.e., an amber STOP codon). An amber mutation results in premature termination of protein synthesis.

amine: Any molecule that has one or more NH_2 groups bound to it.

amino acid: An organic molecule that is the basic building block of protein molecules. It has both an amino group ($-NH_2$) as well as a carboxyl group ($-COOH$) to give this configuration: (H_2N-CH-COOH)-R. The R group represents a side chain that is different in each of the 20 amino acids.

amino acid residue: An amino acid in a polypeptide chain. Each amino acid loses one water molecule when joined to the polypeptide chain.

amino acid sequence: The linear order of amino acid residues in a polypeptide chain or protein.

aminoacyl site (A site): The docking site on the ribosome for the incoming tRNA molecule. The tRNA carries with it the next amino acid to be added to the growing polypeptide chain.

aminoacyl-tRNA synthetase: An enzyme that catalyzes the activation and attachment of a specific amino acid to its specific tRNA molecule. There is at least one of these enzymes for each of the 20 different amino acids.

amniocentesis: A method of obtaining fetal cells which are then tested for potential genetic disorders and chromosomal abnormalities. A sample of amniotic fluid is removed by syringe from the amniotic sac that surrounds the fetus. This fluid contains cells shed from the fetus. Chromosomal spreads (karyotypes) can be made from these cells and checked for abnormalities. The presence of certain chemicals within the amniotic fluid can also signal genetic abnormalities (e.g., high levels of alpha fetoprotein can mean neural tube defects).

Angstrom unit: (Å) A unit of measurement equal to 10^{-10} meters in length. Until recently, it was used when describing unit lengths at the cellular and the molecular level. Most scientific notations have since converted to the use of nanometers (a billionth of a meter or 10^{-9} meters).

antibody: A large Y-shaped protein molecule, produced by certain white blood cells in response to the presence in the bloodstream of a foreign particle such as a bacterial cell or a virus. The antibody that is specific for the invader attaches itself to it and triggers a sequence of biochemical events that destroys the invader, preventing or reducing infection.

anticodon: A set of three nucleotides in a transfer RNA molecule that recognizes the complementary codon on the messenger RNA molecule during protein synthesis.

antiparallel: The opposite polarity of orientation in any nucleic acid duplex. For example, the two strands in double-stranded DNA run in opposite directions. One strand runs from its 5' to its 3' end, and the complementary strand runs from its 3' to its 5' end (i.e., antiparallel to it). This antiparallel arrangement occurs between any two polynucleotide strands that are joined together in a duplex (DNA-DNA, DNA-RNA, or RNA-RNA).

autosome: Any chromosome other than sex chromosomes. For example, in human cells there are a total of 46 chromosomes, which include 22 pairs of autosomes and one pair of sex chromosomes (X and Y chromosomes).

AZT (azidothymidine): A modified form of the nucleotide thymidine monophosphate (thymine, deoxyribose, and phosphate) which is normally incorporated into the DNA molecule during DNA replication. The AZT molecule looks similar enough to thymidine monophosphate that it can be placed into a newly synthesized DNA chain as it is being made by the enzyme reverse transcriptase using the HIV's mRNA template. When AZT is incorporated instead of a genuine thymidine monophosphate nucleotide into the growing DNA strand, the chain cannot be extended any farther. This results in the premature termination of DNA strand synthesis. AZT has been used as a therapy for AIDS patients in the hope of slowing down the replication of HIV particles.

bacteriophage: A virus whose host is a bacterial cell. Bacteriophages (or phages) have been important in the development of molecular genetics. More recently, they have contributed significantly to the development of recombinant DNA technology. Some of the first experiments in gene splicing or "recombinant DNA" were carried out on a phage called "bacteriophage lambda" which infects *Escherichia coli* cells. The phage does not use the middle third of its DNA for the process of reproducing itself inside a bacterial cell. This section can be replaced by foreign DNA from another organism and propagated in future generations of bacteriophage just as if it were a part of the original phage DNA. This finding proved very handy in the early gene-cloning experiments.

base analog: A nitrogenous base that differs only slightly from one of the bases normally found in DNA or RNA (Guanine, Cytosine, Adenine, Thymine, or Uracil). It can be incorporated into DNA or RNA in place of the correct base, increasing the chance of a mutation.

B-DNA: The classical model of DNA elucidated by Watson and Crick. B-DNA is the principal form found within living cells and some viruses. It is a "right-handed" double helix: if you could look down the long axis of the molecule you would see that the two chains wind around each other in a clockwise (or right-handed) direction with 10 nucleotides per 360° turn. The plane surfaces of the base pairs are perpendicular to the long axis of the double helix.

bases of nucleic acids: The nitrogenous bases found in DNA and RNA. There are two classes of bases, the purines and the pyrimidines. In DNA, the purines are Adenine and Guanine, and the pyrimidines are Cytosine and Thymine. In RNA, the purines are also Adenine and Guanine but the pyrimidines are Cytosine and Uracil.

base pair: The linking of two complementary nitrogenous bases by hydrogen bonds within the double helix of DNA.

base pairing rules: See complementary base-pairing rule.

biotechnology: The use of living organisms or cells from living organisms to produce commodities such as proteins and herbicide-resistant plants.

cancer: A disease characterized by the abnormal and uncontrollable growth of cells.

carbon-14 (^{14}C): The radioactive isotope of carbon that is much less abundant than the stable isotope, carbon-12. It emits weak streams of beta particles and has a half-life of 5730 years. In the lab, ^{14}C is incorporated into molecules in place of the normal ^{12}C in order that the movement and ultimate fate of those molecules can be traced. The natural incorporation of Carbon-14 by living things is useful when dating long-dead matter (Carbon dating).

carcinogens: Any physical, chemical or biological agents that cause cancer.

cell: the basic unit of structure and function in all living things.

carrier: Any organism with a diploid genome that carries a detrimental recessive allele in combination with a normal dominant allele at a single locus. The detrimental effects of the recessive allele are hidden by the expression of the dominant allele (which produces the "good protein"). The carrier of the recessive allele may pass the "bad gene" to an offspring who may, by chance, receive a second "bad gene" at the same gene locus from the other parent. When this happens, the offspring will be homozygous recessive at that particular locus and will develop the condition (e.g., sickle-cell anemia or cystic fibrosis).

*centimorgan: A unit named after Dr. Thomas Hunt Morgan, who won the Nobel Prize in 1933 for the chromosome theory of heredity based on his work with the common fruit fly *Drosophila melanogaster*. The centimorgan is a "gene-mapping" unit, and refers to the frequency of crossing over that occurs between any two genes. A centimorgan is defined as the distance along a chromosome within which there is a 1% chance of the occurrence of a crossover. The higher the centimorgan number, the farther apart the two genes are on the chromosome and the more likely they would be separated by a crossing-over event.

central dogma: A theory in genetics that prevailed from the 1950s until 1970, stating that the flow of genetic information was unidirectional and flowed from DNA to RNA to protein. In 1970, however, Dr. David Baltimore and Dr. Howard Temin discovered an enzyme in a class of viruses called retroviruses, which they called reverse transcriptase. This enzyme used RNA to make a DNA strand. This upset the central dogma because genetic information is now known to flow from RNA to DNA.

centromere: A small region on the chromosome where sister chromatids are attached to one another. The centromere is also the attachment site for protein molecules that connect with the fibers of the spindle apparatus that pulls the chromosomes apart to opposite poles of the cell during mitosis and meiosis.

Chargaff's Rule: In 1949, Dr. Erwin Chargaff of Columbia University discovered that in double-stranded DNA, the number of Adenine residues equals the number of Thymine residues, and the number of Guanine residues equals the number of Cytosine residues. Stated another way, the number of purines (A + G) equals the number of pyrimidines (T + C) in DNA. The equation which summarizes Chargaff's Rule is: $(A + G) = (T + C)$.

chiasma: The point of contact between non-sister chromatids of homologous chromosomes during prophase of meiosis. The chiasma is the visible physical evidence for the exchange of genetic material between homologous chromosomes.

chimeric DNA: A term often used to describe recombinant DNA that results from splicing together DNA from two different species (e.g., the splicing of toad DNA and bacterial plasmid DNA).

*chorionic villus sampling: A technique in which a microscalpel is used to snip a very tiny piece of tissue from the fetal part of the placenta. The chorionic villi are tiny finger-like protuberances that the zygote uses to implant itself into the wall of the mother's uterus. The chorion cells have the same genotype as the zygote, so they can be used to make a karyotype, or chromosomal picture. From the karyotype, specialists can determine if there are any genetic defects or chromosomal abnormalities in the growing fetus. Unlike amniocentesis which can be done only after 15 to 18 weeks of gestation, this technique can be performed after only 8 to 12 weeks of gestation, allowing earlier detection of fetal disorders.

chromatid: During the S phase of interphase, the chromosomes appear as very fine threads and the DNA that makes up a large part of each chromosome is duplicated. In prophase of either meiosis or mitosis, the chromosomal threads become densely coiled and appear thicker and shorter. It is evident that they have replicated, and each copy of the original chromosome is called a "chromatid." The chromatids are held together by a centromere.

chromatin: The chromosomes as they appear during interphase of mitosis and meiosis in eukaryotic cells. At this stage, the chromatin is very fine, nearly invisible threads (unlike the densely coiled, rod-like structures that the chromosomes adopt during prophase of both meiosis and mitosis).

chromosome: In prokaryotes, the chromosome is a circular DNA molecule containing the entire set of genetic instructions necessary for the functioning of the cell. In eukaryotes, the chromosomes are rod-like structures composed of a combination of DNA and proteins called histones.

* chromosomal walking: A technique used in molecular biology that involves the use of a genomic library. A genomic library contains the entire nucleotide sequence of the DNA of one complete set of chromosomes of an organism. If we took all of the DNA fragments that were inserted into the cloning vector molecules that make up the library, we would be able to align those DNA inserts from end to end to get the entire linear sequence of genomic DNA for that organism. Since those fragments were initially produced by a restriction enzyme technique called "partial digestion," some of those DNA inserts would have sequences that overlap with many others. What we would end up with on our DNA map is a series of fragments that from left to right would overlap to varying degrees but would ultimately extend farther and farther from left to right until we had the entire linear sequence of genomic DNA for that organism. With the technique of using a "probe" of radioactive DNA to sequentially fish out complementary DNA sequences in either the left or right direction within a genomic library, you are "walking along the chromosome."

* cistron: A stretch of DNA or RNA that codes for

a polypeptide chain, tRNA, or rRNA. In prokaryotes, one long messenger RNA molecule may code for more than one polypeptide chain. In this case, the stretch of DNA that becomes transcribed into the long messenger RNA molecule is said to be polycistronic. The protein molecules coded for by a polycistron are often enzymes that are part of the same metabolic pathway.

cloning: The production of identical copies from an original entity, either DNA molecules from an original molecule or cells from an original cell. The term clone can also refer to a line of living, genetically identical organisms descended from an original organism.

cloning vector: Any vehicle, such as a plasmid or a phage, that can carry DNA to be cloned into a bacteria.

codon: A set of three nucleotides in messenger RNA that codes for 1) a start signal (i.e., initiation of protein synthesis); 2) the placement of an amino acid into a growing polypeptide chain during translation; or 3) a stop signal (i.e., termination of protein synthesis).

complementary base pairing rule: In all nucleotide duplexes, Guanine always pairs with Cytosine. In the DNA double helix, Adenine always pairs with Thymine. When RNA is formed, Adenine pairs with Uracil and Thymine pairs with Adenine. In RNA duplexes, Adenine always pairs with Uracil.

complementary DNA: (also known as cDNA) DNA made from a messenger RNA molecule by the action of the enzyme reverse transcriptase. In the lab, the first step is to isolate the mRNA from the cells. Then the enzyme reverse transcriptase uses the single stranded mRNA molecule as a template to synthesize a complementary single strand of DNA. The RNA template is removed, releasing the single-stranded DNA molecule. With the enzyme DNA polymerase and appropriate deoxyribonucleoside triphosphates, this single strand of DNA serves as a template for the formation of its complementary sister strand to result in a double-stranded DNA molecule. Since the splicing-out of any introns has already taken place within the mRNA molecule (i.e., the mRNA is mature), the cDNA will also have no introns within its sequence.

cDNA library: A collection of cloning vectors, such as plasmids or lambda bacteriophages, containing fragments of DNA complementary to messenger RNA molecules extracted from a tissue sample. The cDNA molecules are obtained by the action of the enzyme reverse transcriptase on messenger RNA templates, so the cDNA library represents only those genes that were actively expressed in the nuclear DNA (i.e., transcribed into mRNA). There is one main difference between a genomic DNA library and a cDNA library. A genomic DNA library is a collection of DNA fragments that represent the entire nucleotide sequence, and hence all of the genes as well as the regulatory regions and introns, of the organism under study. The cDNA library includes only those genes that are actually transcribed in the nucleus. (Note that the cDNA library is without the regulatory regions such as promoters, enhancers, or introns that are present in the genomic library). This also means that the cDNA library represents only those genes that were active in the cell from which the mRNA was obtained. For example, all cells of an organism have the same genomic DNA in their nucleus; however different cell types (e.g., brain cells vs. liver cells) have different types of messenger RNA in them. Brain cells will have mRNA that codes for brain cell receptor proteins while liver cells will have mRNA that codes for proteins that occur only in liver cells.

cDNA clone: A double-stranded DNA molecule, complementary to a particular messenger RNA molecule, spliced to an appropriate cloning vector.

* conserved sequence: A sequence of nucleotides in a gene or a sequence of amino acids in a polypeptide chain that has had few or no changes over significantly long periods of time through evolution. These sequences often control vital life functions.

* core DNA: The length of DNA that wraps around the octet of histone proteins in a nucleosome.

---

cosmid: A cloning vector designed for cloning large fragments of DNA. The name cosmid is derived from the fact that the vector is a plasmid on which "cos" sites have been inserted. A cosmid is a handy cloning vector because it can be used to clone large eukaryotic DNA fragments of about 45 000 base pairs long, as opposed to plasmid cloning vectors that can only handle DNA inserts of up to about 10 000 base pairs long.

* cos sites: Nucleotide sequences of phage lambda DNA that are recognized by the proteins that facilitate the act of packaging the approximately 50 000 base pairs of the phage DNA into the phage's protein capsule head. In lambda phage particles, the DNA is in the form of a linear double-stranded molecule with single-stranded complementary ends (these are called cohesive ends). Each of these cohesive ends or single-stranded cos sites are only 12 nucleotides long.

crossing over: The process of homologous chromosomes coming into physical contact with one another during meiosis (i.e., in sex-cell division) and exchanging sections of chromosomes with one another. Crossing over results in greater genetic diversity in the gametes, which in turn results in greater genetic diversity in the offspring.

* cryptic gene: As its name implies, this is a gene that lies mysteriously silent because of a single nucleotide substitution. It can, however, be re-activated by a single mutation at the original substitution site.

cytosine: One of the four nitrogenous bases, found in DNA and RNA.

deoxyribonucleic acid: The nucleic acid containing the genetic information found in the cells of all organisms. It is a long molecule made up of two complementary polynucleotide chains that wrap around each other to form a double helix. DNA molecules are the largest biologically active molecules known with molecular weights of up to 1×10^8 Daltons.

deoxyribonucleoside triphosphate: Molecules made up of a nucleotide plus two extra phosphates. During DNA replication and transcription, these molecules are incorporated into the growing DNA or RNA strand, and the extra two phosphates are split off.

deoxyribose: The 5-carbon sugar molecule that is a component of DNA. It differs from its close relative, ribose sugar, in that it does not have an oxygen molecule at the 2-carbon position.

disulfide linkage: The amino acid cysteine has a sulfur atom at the terminal end of the side chain. This sulfur atom is capable of participating in a covalent bond with the terminal sulfur atom of another, nearby cysteine residue within the same protein chain or between different protein molecules. The disulfide linkages between cysteine residues in alpha keratin (the protein that makes up hair) determine the shape of hair.

DNase (Deoxyribonuclease): A class of enzymes that help to hydrolyze the phosphodiester linkage between adjacent nucleotides in a DNA molecule. It cuts the DNA molecule into smaller fragments.

* DNA clock hypothesis: A concept that can be used to determine approximately when two different species diverged from a common ancestral organism. Differences in the nucleotide sequences of their respective DNAs are analyzed. One common example of this type of "ancestral-tree" analysis is the comparison of the amino acid sequence of a respiratory enzyme called cytochrome c that occurs within the mitochondria of all eukaryotic cells and some prokaryotic cells. The theory suggests that the shorter the period since two species diverged on the evolutionary tree, the fewer the differences between the two amino acid sequences of cytochrome c. Since the amino acid differences reflect the differences in the nucleotide sequence of the DNA that coded for those amino acid sequences, the degree of divergence in the DNA sequences between two species can be used to estimate their divergence point in evolution. The cytochrome c molecules in chimpanzees and dogfish differ by 24 amino acids, in chimpanzees and horses differ by 11 amino acids, in chimpanzees and Rhesus monkeys differ by only one amino acid, and in chimpanzees and humans are identical.

DNA fingerprint technique: A technique, also known as DNA profiling, invented by Professor Alec Jeffreys of Leicester University in England in the early 1970s. He discovered "tandem repeat sequences" of nucleotides scattered throughout the genome; these serve as the basis of DNA fingerprinting. Restriction endonuclease sites are also scattered throughout the human genome. The distance between two sites is often affected by the number of tandem repeat sequences that happen to lie between the sites cut by a restriction endonuclease. The lengths of the fragments formed by any one enzyme vary widely between any two individuals. If you cut up the DNA from any two people using the same enzyme (or the same combination of enzymes), you get a large population of different length fragments of DNA called "restriction fragment length polymorphisms" or RFLPs for short. The set of RFLPs from person "A" differ in their size distribution from the set of RFLPs from person "B." The trick highlighted by Dr. Jeffreys was that each of these tandem repeat sequences share a common 10-15 base-pair sequence of nucleotides. Therefore, a radioactive DNA probe containing this core sequence will hybridize to many of the RFLPs because they contain this 10-15 base-pair sequence. The "banding pattern" (which resembles a vertical barcode) that results from the electrophoresis gel separation of the DNA fragments is different for each person (except for identical twins). In criminal investigations, DNA fingerprinting is used to identify the guilty party from hair, blood, semen, saliva, or skin tissue left by the perpetrator at the scene of the crime. Like magic, the polymerase chain reaction can take minute amounts of DNA left at a crime scene and make enough DNA for a DNA fingerprint test.

* DNA grooves: The DNA molecule has two grooves that follow the path of the winding double helix, a major groove which is 1.2 nanometers wide and a minor groove which is 0.6 nm wide. The major groove is a little deeper than the minor groove. Atoms along the inner walls of each groove can interact with incoming protein molecules that recognize particular DNA sequences.

One example is DNase (deoxyribonuclease) that binds preferentially to the minor groove of the DNA double helix.

DNA gyrase: An enzyme that acts as a molecular "swivel" to unwind the closed circular double-stranded DNA molecule of bacteria during DNA replication. It is classified as a type of enzyme called a "topoisomerase." During DNA replication, a helicase enzyme breaks the hydrogen bonds between the complementary bases to unwind the double helix. Because one chain is wound around the other, unwinding at any point will cause knotting and supertwisting in other regions of the duplex. DNA gyrase makes tiny breaks in the sugar-phosphate backbone to relieve this "supercoiling" and then reseals the nicks.

DNA helicase: A class of enzymes in bacteria that unwinds the double helix just ahead of the enzyme DNA polymerase III during DNA replication. Driven by the hydrolysis of ATP, it unwinds the double helix by breaking the hydrogen bonds between complementary bases. (The rep protein is a DNA helicase, found in *E. coli* cells.)

DNA library: A large collection of recombinant DNA molecules made up of DNA fragments from a single type of organism, spliced to cloning vectors. The complete collection of these recombinant DNA molecules contains the complete nucleotide sequence of the genome of the organism that the DNA fragments came from.

DNA ligase: An enzyme that catalyzes the sealing of single-stranded nicks in the DNA sugar-phosphate backbone by catalyzing the formation of a 3' – 5' phosphodiester linkage between the 3' end of one DNA fragment and the 5' end of another. This enzyme is very useful in gene splicing experiments. When two fragments of DNA from different species have complementary "sticky ends," the bases of the complementary single-stranded ends form hydrogen bonds together. DNA ligase completes the splicing job by sealing the open nicks in the sugar-phosphate backbone at the splice points. The DNA ligase used in genetic engineering techniques is called T-4 DNA ligase and is derived from a bacteriophage called T-4 that infects *E. coli* cells. T-4 DNA ligase has the ability to seal the nicks in double-stranded DNA as well as the ability to join two DNA fragments with ends that are completely base-paired (i.e., blunt ends). This makes it extremely useful when making recombinant DNA molecules.

* **DNA methylation:** The addition of methyl groups (CH_3) to certain nitrogenous bases within the DNA double helix after DNA replication. This protects the DNA from the cell's own restriction enzyme cutting action. DNA methylation also contributes to the control of gene expression.

DNA polymerase: Also known as the Kornberg enzyme, DNA polymerase I is named after Arthur Kornberg who won the Nobel Prize for its discovery in 1955. In *E. coli* cells, this enzyme catalyzes the formation of a new single strand of DNA using deoxyribonucleotide triphosphates (dNTP) and an already existing single strand of DNA as a template. The enzyme attaches an incoming dNTP to the 3' end of the growing DNA single strand. Therefore, the direction of growth of the new DNA chain is from the 5'-end to the 3'-end. Soon after, it was discovered that two other types of DNA polymerases exist (DNA pol II and DNA pol III). In *E. coli*, DNA pol III handles the major function of DNA replication. All DNA polymerases are now known to have similar action.

DNA probe: Typically a small segment of DNA that is chemically synthesized in the lab. This small stretch of DNA is made to order and radioactively labeled with Phosphorus-32. The "hot" DNA is used to hybridize, by hydrogen-bonding, to a complementary sequence in a gene of interest in genomic DNA (hence the name "probe"). Because the DNA is radioactively labeled, the Southern Blot analysis will show exactly where the probe hybridized to the DNA under study. Therefore, the location of that gene is indicated (by a dark band on an X-ray film) within the gel separation of the different lengths of DNA fragments.

* **DNA replicase system:** The set of enzymes and specialized proteins necessary for DNA replication within a cell. Also referred to collectively as the replisome. In *E. coli* this complete set is comprised of about 20 different enzymes and proteins, each one with a specific job.

* **DNA supercoiling:** The coiling and twisting of the double helix of DNA in much the same manner that an elastic band will supercoil if you fix one end of the elastic and twist the other end. The result is a tightly twisted and knotted appearance. A supercoil is also called a superhelix.

dominant allele: An allele that is expressed in an organism whether it is in the homozygous or the heterozygous state.

double helix: The natural configuration of two complementary DNA strands wound around each other in an anti-parallel fashion and in a clockwise turning direction.

Down syndrome: A human condition resulting from trisomy of chromosome #21 (i.e., an extra chromosome #21). The disorder is characterized by heart and respiratory ailments as well as mental retardation.

electrophoresis: A technique in which a sample containing a mixture of molecules is put on an inert support and then the whole system is exposed to an electric field. Charged molecules move in the electric field and are thus separated on the basis of their charge and size. Porous gels are often used as a support for DNA fragments. Fragments of DNA are pipetted onto one end of the gel and then an electric current is passed through the buffer solution in which the gel is immersed. Because DNA has an overall negative electric charge, the fragments migrate through the gel away from a negative electrode toward a positive electrode. After the current has run through the gel for a few hours, the different sized DNA fragments have migrated to different distances along the gel. (i.e., the shorter fragments have migrated farther along the gel than the longer fragments.) You end up with a nice separation of DNA fragments according to the number of nucleotides in the particular fragment. Agarose gel and polyacrylamide gel electrophoresis are standard techniques used in many procedures including restriction fragment length polymorphism (RFLP) analysis and DNA sequencing respectively.

* enhancer: A sequence of nucleotides in a DNA molecule that can powerfully increase the rate of transcription of a particular gene. The enhancer element can be on either side of the gene it affects and can be quite far away from the gene (as much as thousands of base pairs away from the promoter of that gene). The enhancer effect is thought to work by serving as a binding site for certain sequence-specific DNA-binding proteins. Once the DNA-binding protein attaches to the enhancer sequence, it loops out the DNA that spans the distance between the enhancer sequence and the promoter of the gene that it affects. This brings the enhancer sequence into physical contact with the promoter region of its gene. This loop-structure in some way makes it easier for the RNA polymerase molecules to bind to the promoter and begin transcription of that gene.

enzymes: Biological catalysts. Enzymes lower the activation-energy level required by biochemical reactions so that they may proceed quickly at lower temperatures (such as body temperature in humans). They differ from inorganic catalysts in their extreme specificity for a particular substrate molecule. Most enzymes are proteins, but it was discovered in 1982 by Thomas Cech that ribozymes (RNA molecules that act as enzymes) can catalyze the splicing out of introns without any protein intervention.

Escherichia coli: *E. coli* is a single-celled, rod-like bacteria which exists within the small intestines of many vertebrates. It is a prokaryote (without a nuclear membrane) and it is the most closely studied and well-understood organism on earth. It contains a single chromosome, a double-stranded molecule of DNA about 4.7 million base pairs long. Most of our understanding of microbiology and biochemistry comes from experimentation done on this bacterial cell. It was the workhorse of the early genetic engineering experiments because it serves as a good host for a wide variety of viral, plasmid, and cosmid cloning vectors.

essential amino acids: Bacteria and plants can synthesize all 20 of the amino acids used to make proteins. However, many animals lack the enzymes necessary for synthesizing some of the necessary amino acids. The amino acids that a given species cannot synthesize but must obtain from its diet are called the "essential amino acids" for that species. For example, the following eight amino acids are essential for humans: isoleucine, leucine, lysine, methionine, phenylalanine, threonine, tryptophan, and valine.

eukaryotic cell: A cell that possesses a membrane-enclosed nucleus, chromosomes built of DNA and protein, a cytoskeleton, and membrane-bound organelles. All multi-cellular organisms are eukaryotes, as well as single-celled organisms other than bacteria and cyanobacteria. (As distinguished from prokaryotic cell.)

eugenics: The attempt to improve characteristics of future generations of humans by the selection of mating partners based on what are perceived as being superior traits.

exon: A nucleotide sequence, part of a split gene in eukaryotes, that is transcribed into messenger RNA and survives the splicing out of intervening sequences (introns) during the processing that produces mature mRNA. The exons are the regions of DNA that actually carry the code for the eventual placement of amino acids into a polypeptide chain.

exonuclease: An enzyme that digests DNA by chewing away nucleotides at the terminal ends of the strands.

* expression vector: A recombinant DNA molecule that is constructed especially to aid the production of a desired protein molecule from a specific gene. If, for example, a foreign gene without its own regulatory elements were spliced into a plasmid DNA molecule, the foreign gene's regulatory elements would not be compatible with the gene expression machinery of the host bacterial cell. Consequently, the original regulatory element is removed before splicing. Then a DNA fragment carrying a strong promoter and protein synthesizing regulatory elements that are compatible with the bacterial cell is spliced into the recombinant DNA molecule just ahead of the gene of interest. The result is that the host bacterial cell starts to express the foreign gene and begins to produce large amounts of the desired protein.

fermentation: The chemical breakdown of an organic molecule such as sugar in the absence of oxygen, yielding energy and other by-product molecules such as lactic acid in yogurt and ethanol in alcoholic beverages.

frame shift mutation: A mutation that involves the deletion or addition of one or more nucleotides within a DNA molecule. Since the nucleotide triplets are read as a series of three nucleotides at a time, every nucleotide triplet after the mutation will be changed because every nucleotide after the frame shift mutation will be either pulled back one position (in the case of a single nucleotide deletion) or bumped forward one position (as in the case of an addition of a single nucleotide). Therefore, such a mutation results in a "shift of the reading frame" of the nucleotide triplets that lie after the location of the deletion or addition. This will result in a different sequence of amino acids in the protein chain from the point of the mutation onward, which usually means that the protein is nonfunctional.

gamete: A reproductive sex cell (an egg or a sperm cell). Gametes have half the chromosome number of the body cells (somatic cells).

guanine: One of the four nitrogenous bases, found in DNA and RNA.

gene: The fundamental unit of inheritance. A section of a nucleic acid (DNA, and RNA in some viruses), a gene carries within its sequence of nucleotides the information required for 1) the sequential attachment of a series of amino acids to form a polypeptide chain; 2) the construction of a messenger RNA, transfer RNA, or ribosomal RNA molecule; or 3) the regulation of transcription of these sequences.

gene cloning: Multiplication of any DNA segment representing a gene using an appropriate cloning vector such as a plasmid or bacteriophage DNA molecule. This actual multiplication of the gene + vector DNA is carried out in a bacteria cell such as *E. coli* by its biochemical machinery.

gene locus: The physical location of a particular gene along the length of a chromosome.

gene splicing: The technique used to join together separate segments of DNA from different sources (often from different species).

gene therapy: The genetic engineering process whereby a normally functioning gene is inserted into the genome of an organism with an inherited condition. The newly inserted gene would produce a normally functioning protein. This technique is still in the experimental stage.

genetic code: See universality of the genetic code.

genetic code dictionary: A list of all 64 codons (nucleotide triplets) that code for the 20 amino acids as well as the start and stop signals. (See page 29).

genetic linkage map: A map of the relative positions of genes on a chromosome that is deduced by how often these genes are inherited together (genetic recombination experiments).

genetics: The study of the inheritance of genes and the physical traits that arise from them.

genetic screening: The testing of cells from an individual (frequently fetal cells) to detect the presence of alleles carrying certain genetic disorders.

genome: All of the genes carried by one complete set of non-homologous chromosomes, or, in other words, all of the genes within a single sex cell. In humans, for example, the genome is the genes that occupy all of the loci on the 23 chromosomes.

genomic library: A collection of DNA fragments that were produced by cutting genomic DNA with restriction enzymes (or produced by ultrasonic-shearing of the genomic DNA) and splicing them to appropriate cloning vector molecules such as plasmid or bacteriophage DNA. The genomic library will contain enough DNA inserts from the organism under study to represent the organism's entire genome.

genotype: The alleles that an individual carries for one or a few genes. For example, a brown-eyed person may carry an allele for brown eyes (B) and one for blue eyes (b). The individual's genotype for eye color would be "Bb." Genotype can also apply to all the alleles of all of the genes in the organism.

germ-line gene therapy: The genetic engineering process whereby normal genes are inserted into the chromosomes of a fertilized egg in an attempt to improve the genome of the developing organism. It is important to note that if a new gene is inserted into the chromosome structure of a fertilized egg, then every single cell that arises from that zygote will contain the new gene. This includes the sex cells (i.e., germ-line cells) of the organism. When the organism reaches sexual maturity, it will pass on the new gene to future generations. Technology for human germ-line gene therapy does not yet exist. However, it may be developed in the future.

half-life: the amount of time required for half of the mass of a radioactive element to decay by emitting alpha or beta particles or gamma rays.

haploid cell: A cell that contains only one set of chromosomes of a given organism. For example, the sex cells (sperm and egg) of mammalian organisms are haploid cells which contain only half the chromosome number of the somatic cells (body cells).

helicase: A class of enzymes that catalyze the breaking of hydrogen bonds between the nitrogenous bases of the two strands of DNA, separating them in order that DNA replication can take place.

helper T cells: A class of white blood cells that are a part of the mammalian immune system. They interact with other types of white blood cells in the immune system called B cells, killer T cells, and macrophages. Helper T cells "dock" with the B cells and stimulate them to produce specific antibodies to inactivate a foreign invader. They also dock with macrophages and killer T cells, stimulating them to attack invaders and to secrete proteins called lymphokines (interferons and interleukins) which stimulate other cells of the immune system. The helper T cells are of great significance when discussing the mechanism of AIDS development because the HIV particle has a strong attachment affinity for a protein (called CD4) on the surface of the helper T cell. Once the HIV particle binds to the surface of the helper T cell, it can infect it and eventually cause its death. The helper T cells play such a key role in the strength of the immune system that if they are knocked out by HIV, the infected person's immune system wears down and the person succumbs to other infectious agents such as pneumococcal bacteria that cause pneumonia.

hemoglobin: The iron-containing pigment found in red blood cells that attaches to oxygen and carries it to the various parts of the body. Hemoglobin is composed of four polypeptide chains, each one of which is folded around an iron-containing "heme" group. Two chains are called "alpha" chains and the other two are called "beta" chains. In humans, each of the two alpha chains contain 141 amino acids and the two beta chains each contain 146 amino acids. The hemoglobin molecule has a molecular weight of 64 500 daltons. In humans, the genes for the two beta chains reside on chromosome #11 and the genes for the two alpha chains are found on chromosome #16.

heredity: The passing on of genes and corresponding physical traits from one generation to the next.

heterozygote: An organism that has two different forms of a gene (alleles) for a particular trait. For example, if a person has one gene for blue eye color and one for brown eye color, the person's eyes are brown and he or she is classified as a heterozygote for eye color.

histones: A group of proteins rich in basic amino acids that bind to the DNA of eukaryotic organisms. There are five different types of histone proteins, classified according to the amounts of the amino acids lysine and arginine that they contain. The five different types of histones are called H1, H2A, H2B, H3, and H4. The histones are believed to play a role in the compacting of DNA into a smaller space in the nucleus as well as having a key role in regulating gene expression.

HIV: The acronym for Human Immunodeficiency Virus, believed to cause AIDS. HIV belongs to a special class of viruses called "retroviruses" which contain RNA as their genetic material. This viral

RNA directs the eventual synthesis of a double-stranded DNA molecule which becomes integrated into one of the host cell's own chromosomes. This foreign DNA insert can do one of two things: lie dormant or become actively expressed to produce many more HIV particles within the cell. Two types of HIV are known. HIV-1 describes the virus most commonly found to cause AIDS in North America and Europe. HIV-2, a closely related but distinctly different form of the virus, has been isolated from infected people in West Africa.

* Hogness box: A small region within eukaryotic DNA that serves as the binding site for the enzyme RNA polymerase II. (It was named after the discoverer Dr. David Hogness). RNA polymerase II is the enzyme responsible for transcription of the DNA template strand into a complementary messenger RNA molecule. The Hogness box is located along the DNA molecule about 30 base pairs upstream (i.e., to the 5' side) from the transcription start site of eukaryotic structural genes and is 7 base pairs long. The nucleotides that make up the Hogness box are TATAAAA (from the 5' to 3' direction along the DNA coding strand). For this reason the Hogness box is also referred to as the "TATA box" of eukaryotic DNA.

(CAT Box)	(TATA Box or Hogness Box)	
5'- - GGCCAATCT- - - -	TATAAAA- - - -	Transcription Start Site
-50 to -100 base pairs	-30 base pairs	0 point

About 50 to 100 base pairs upstream (i.e., to the 5' side) from the transcription start site is a sequence of nine nucleotides with CCAAT in the central region of the 9mer. This "CAT Box" also plays a role in the binding of RNA polymerase to the promoter region of eukaryotic DNA. (See also Pribnow Box.)

* homeobox: A highly conserved DNA sequence that plays a role in the patterns of differentiation and development of eukaryotic organisms. It is 180 base pairs long and codes for a polypeptide chain of 60 amino acids. The polypeptide constitutes a key part of a class of proteins that regulate gene expression by binding to strategic genes and turning them on or off. The homeobox was first discovered in the genes of the common fruitfly, *Drosophila melanogaster* by Dr. Edward Lewis at the California Institute of Technology in the late 1970s. The same homeobox stretch of 180 nucleotides is found in many genes that have been shown to control morphological development. Collectively they are called "the homeotic genes." (See conserved sequence.)

* homeotic genes: A class of genes that control and co-ordinate the overall design of an organism by coding for proteins that dictate the ultimate position and developmental fate of groups of cells in the embryo. Homeotic genes are thought to behave as "master-switches" that control many other genes involved in the orchestration of development.

homologous chromosomes: Chromosomes that pair up along the center plate of the sex cell during meiosis I. Each chromosome of the homologous pair has the same length, centromere position, stain-banding pattern, and linear order of genes at the same loci. One chromosome of each pair of homologous chromosomes came originally from the mother of the organism and the other came from the father.

homozygote: An organism with two identical alleles at a given gene locus on two homologous chromosomes. For example, if a person has two genes for blue eye color, then he or she has blue eyes and is classified as being homozygous for eye color.

hormone: Any organic molecule that is produced in one part of an organism and is transported to a "target site" where it affects the target site's function.

hydrogen bond: A weak, non-covalent chemical bond which arises between a slightly positively charged hydrogen atom and a slightly negatively charged nitrogen, sulfur, or oxygen atom. These slightly charged atoms may be in the same molecule or in different molecules. Hydrogen bonds occur between the complementary nitrogenous bases, holding the two strands of the DNA double helix together.

immunoglobulins: A category of proteins produced by a class of white blood cells called B-lymphocytes. Immunoglobulins are Y-shaped molecules made up of four polypeptide chains (two short chains called "light chains" and two longer chains called "heavy chains"). They are secreted into the bloodstream by mature B-lymphocytes (also called plasma cells) in response to an invasion of foreign particles or proteins into the bloodstream. The immunoglobulin molecules attach themselves to specific regions on the surface of the invader. This marks the invader for destruction by another type of immune-system cell, or prevents the invader from attaching itself to a landing site on a healthy cell. There are five classes of immunoglobulins, named according to the atomic composition of the "stem" part of the Y-shaped molecule. (When referring to any immunoglobulin, the abbreviation "Ig" is used in place of the word immunoglobulin). The five types of immunoglobulins are: IgG, IgA, IgM, IgE, and IgD. IgM is the first class of antibody molecule to appear in the bloodstream in response to an invader. IgG is the most common antibody found in blood serum. IgA is associated with resistance to infectious diseases of the respiratory and digestive tracts, while IgE is associated with the allergic response. IgD plays a role in activating B cells.

* inducers: Molecules that interact with the protein molecules that regulate gene expression. Inducers can stimulate the production of large amounts of particular enzymes. (For example, the milk sugar (lactose) is an inducer that stimulates the production of lactase, the enzyme that breaks it down into its two components, glucose and galactose.

initiation codon: (Also known as the "start codon.") The nucleotide triplet AUG in messenger RNA that codes for the first amino acid of a polypeptide chain. In prokaryotes, the first amino acid is N-formyl methionine, and in eukaryotes it is just plain methionine. The codon AUG also codes for methionine residues in internal positions within polypeptide chains. One might ask, "If AUG is the start codon for all protein synthesis of all organisms, then why don't all finished proteins start with the amino acid N-formyl methionine in bacteria and methionine in eukaryotes?" The reason for this is because of a process called "post translational modification" of the polypeptide chain. Immediately after the polypeptide chain has been synthesized, enzymes may chop off the ends by varying amounts, depending on the function of that particular protein.

* initiation factors: Proteins required to start protein synthesis. Initiation factors interact with the ribosomal subunits as well as the messenger RNA and transfer RNA molecules to form the initiation complex which starts protein synthesis.

* intercalating mutagen: Any flat molecule that can insert itself between two nucleotides in the DNA molecule, resulting in a frame shift in the reading frame of the codons after the site of insertion. Two examples of DNA-intercalating mutagens are proflavin and ethidium bromide.

intron: A stretch of non-coding nucleotides that are part of a split gene in eukaryotic DNA. After introns are transcribed in the messenger RNA, they are spliced out and degraded as the precursor mRNA is processed into mature mRNA molecules. Most genes in eukaryotes contain introns, as do the genes within the DNA of mitochondria and some chloroplasts. The length of introns can vary from less than 100 to more than 10 000 nucleotides. There is little sequence similarity between introns except for the regions at the ends. These boundary regions of the introns act as recognition sites for the enzymes that cut the introns out and re-splice the remaining mRNA ends together. Introns have not been found in the DNA of any prokaryotes to date.

in vitro: A term that means "outside of a living system." Another way of putting it would be to say "in a test tube" as in the term "*in vitro* fertilization."

in vivo: A term that means "within a living system" (i.e., within a cell or a whole organism).

* jumping genes: See transposons.

*junk DNA: Non-coding, highly repetitive, functionless stretches of DNA that have been retained within the genome of a eukaryotic organism over many millions of years. No one knows why junk DNA has been retained, but at least some of it does serve a useful purpose. Junk DNA is concentrated in the region of the centromere of the chromosome.

karyotype: A picture showing the number, sizes, and shapes of the metaphase stage chromosomes of an organism. Karyotyping is a standard procedure used to detect chromosomal abnormalities in a growing fetus. First, some fetal cells are taken from the mother by amniocentesis or chorionic villus sampling. Then the cells are grown and chemically treated in a way that releases all 46 chromosomes which are then stained with a dye and spread out for visual inspection. In this way, any gross aberrations in the number or size of the chromosomes can be detected before birth.

keratin: A structural protein that is a major component of hair, nails, wool, and horns of mammals and feathers of birds.

lagging strand of DNA: during DNA replication, only one newly-formed strand (the leading strand) is synthesized continuously in the same direction as the unwinding of the double helix. The other newly formed strand (the lagging strand) is synthesized in short segments away from the replication fork. This is because the synthesis of a new DNA strand occurs in the 5' to 3' direction from a DNA template that runs in the 3' to 5' direction, and the two strands of the original DNA duplex are in an antiparallel orientation. The breaks in the sugar-phosphate backbone of the discontinuous fragments of the lagging strand are later sealed by DNA ligase. These fragments of the DNA lagging strand are often called "Okazaki fragments" after the scientist that discovered them.

* leader sequence of mRNA: A sequence of ribonucleotides on the messenger RNA molecule that do not get translated into amino acids. The leader sequence stretches from the 5' end of the mRNA molecule to its start codon (AUG). The leader sequence is believed to contain regulatory signals such as where the ribosomal subunits should bind to the mRNA for translation into protein.

leading strand of DNA: In DNA replication, the newly formed chain of DNA that is synthesized continuously without interruption in the 5' to 3' direction. The DNA template strand, used for making the new leading strand, is in the 3' to 5' orientation.

* leucine zipper: A sequence of amino acids in one of the four major classes of DNA-binding proteins that control gene transcription. The leucine zipper is about 30 amino acids long with a repeat of a leucine every 7 amino acid residues. This region of the protein forms an alpha-helix. Two polypeptide chains, each with a leucine-rich alpha-helix, bind together in a "zipper-like" formation to form the shape of the active DNA binding protein. This protein then binds to the DNA in regions that regulate gene transcription into mRNA. (The other three major classes of DNA-binding proteins are called "helix-turn-helix," "helix-loop-helix" and "zinc finger" proteins.)

linkage group: All of the genes located on the same chromosome.

locus: The actual position that a gene occupies along the length of a chromosome.

marker: A gene that when expressed confers upon the organism some easily identifiable physical characteristic. An example of this is the gene for antibiotic resistance of certain plasmid molecules that exist within bacteria. The plasmid called pBR322 has two genetic markers which are genes that give resistance to the bacteria to the antibiotics ampicillin and tetracycline. Another type of marker is a readily identifiable section of a chromosome whose inheritance pattern can be tracked (such as RFLP markers and restriction enzyme cutting sites.)

meiosis: The process of division of diploid cells that results in the production of haploid cells, which have half the chromosome number of the original cell. In meiosis, a single parent cell gives rise to four daughter cells (each with half the chromosome number of the parent cell).

messenger RNA: (mRNA) One of the three types of RNA (the other two are ribosomal RNA and transfer RNA) synthesized from a DNA template. The sequence of nucleotides on messenger RNA determines the sequence of amino acids in a polypeptide chain. Each set of three nucleotides on the mRNA is called a "codon" and codes for one amino acid. In eukaryotes, precursor mRNA is synthesized in the nucleus. It then undergoes the cutting out of introns and resplicing to bring the exons into continuity. The mature mRNA then moves from the nucleus to the ribosomes in the cytoplasm. In prokaryotes, mRNA contains no introns, only exons.

* methylated guanosine cap: One of the modifications that eukaryotic mRNA undergoes in addition to the removal of the introns. A guanine base with a methyl group ($-CH_3$) bound to it is attached to the 5' end of the mRNA by an unusual 5' to 5' linkage. This modification is also called a 5' cap and occurs very shortly after the initiation of transcription and before the splicing out of the introns. The 5' cap is believed to protect the mRNA against digestion by nucleases and to provide a binding site for the ribosomes. 5' capping does not occur in prokaryotes.

mitochondria: Organelles found in the cytoplasm of eukaryotic cells. They are the sites of aerobic respiration that yield the energy-containing ATP molecules. They contain some circular double-stranded DNA, but also rely on nuclear genes. The strong similarity between mitochondria and bacteria led to the theory that mitochondria were once a free-living form of aerobic (oxygen-using) bacteria. They were taken up by the very primitive (anaerobic) eukaryotic cells and harbored in the cytoplasm. The two cells existed as symbiotic partners, the "bacteria" producing more energy than the eukaryotic cell alone and the eukaryotic cell cytoplasm providing an excellent environment for the "bacteria."

mitochondrial DNA: A circular double-stranded DNA molecule contained in mitochondria (about 5 to 10 copies per mitochondria). In mammals, the mitochondrial DNA is usually about 30 000 base pairs long. The genetic code of the mitochondrial DNA of some organisms (e.g., yeast, paramecium, mycoplasma and mammals) differs slightly from the universal genetic code. (See universality of the genetic code.)

mitosis: The division of a cell (not a sex cell) to produce two identical daughter cells, each with the same number of chromosomes as the parent cell. There are four stages in mitosis: prophase, metaphase, anaphase, and telophase (PMAT for short).

* monoclonal antibodies: Antibodies derived in the laboratory from a clone of identical antibody-producing cells of the immune system. Monoclonal antibody molecules are all identical in structure and type, and in their specificity for a certain type of antigen. Monoclonal antibodies come from a cell that is actually a hybrid formed by the fusion of a cancer cell, called a myeloma, with a white blood cell, called a B-lymphocyte, that has matured into an antibody-secreting plasma cell. This fusion of the myeloma cell with the plasma cell is called a "hybridoma." The myeloma cell, which never stops dividing (a feature of all cancer cells), contributes immortality to the hybridoma while the B cell contributes the antibody production. You end up with a kind of hybrid cell that continuously pumps out large amounts of pure antibody of a certain specificity. This revolutionary procedure was developed by Dr. George Koehler and Dr. Cesar Milstein in 1975. In 1984, they were awarded the Nobel Prize for the hybridoma technique of monoclonal antibody production.

* multigene family: A set of genes that arises by duplication and variation from an ancestral gene. Genes of a multigene family may be grouped together on the same chromosome or spread out among different chromosomes. Some examples of multigene families are those genes that encode the different types of hemoglobins, histones, immunoglobulins (antibodies) and the proteins of the major histocompatibility complex in mammals (proteins on the surface of white blood cells that recognize "self" from "non-self" and cause organ transplant rejections).

mutagen: A substance that causes mutations to occur within DNA.

mutagenesis: The making of a mutation.

mutant: A gene that has been changed or an organism that has experienced a change in its genetic makeup.

mutation: A change in the nucleotide sequence of DNA that can be inherited from one generation to the next.

nitrogen-15 (^{15}N): The radioactive isotope of nitrogen that is much less abundant than the stable isotope, nitrogen-14. In the lab, it is incorporated into biomolecules for the purpose of tracking their path and ultimate fate in biochemical reactions.

non-disjunction: The failure of homologous chromosomes to segregate properly and move to opposite poles during meiosis. This results in one daughter cell getting both of the chromosomes that experienced non-disjunction while another daughter cell gets none. Non-disjunction is the cause of human disorders such as Down's syndrome (trisomy 21) and XXY, XO, and XYY people.

nonsense codon: Another term for a stop codon.

nonsense mutation: A mutation that converts a codon that originally codes for an amino acid to a codon that terminates chain elongation (i.e., a stop codon: in messenger RNA, either UAA, UAG, or UGA).

nuclease: Any enzyme that digests nucleic acids.

nucleic acid: A polymer of nucleotides. DNA is a polymer of deoxyribonucleotides and RNA is a polymer of ribonucleotides.

nucleoprotein: A combination of nucleic acid and protein. Two main classes of basic proteins (as opposed to acidic) bind to DNA. Protamines have low molecular weights and histones have higher molecular weights. Protamines replace histones in the DNA of sperm cells of many vertebrate species. The positive charges of these basic proteins are attracted to the negative charges of the phosphate groups in the backbone of the DNA double helix, resulting in a neutrally charged complex.

nucleoside: The combination of a nitrogenous base and a pentose sugar. The nucleosides associated with RNA are adenosine, cytidine, guanosine, and uridine. The nucleosides associated with DNA are deoxyadenosine, deoxycytidine, deoxyguanosine, and deoxythymidine.

nucleosome: A bead-like structure, found only in eukaryotes, that is a combination of DNA wrapped around an octet of histone proteins. More specifically, the nucleosome consists of about 145 base pairs of double-stranded DNA wrapped about one and three-quarter times around a core of eight histones. Nucleosomes are repeating units along the double helix, separated by about 60 base pairs of DNA. This structural configuration results in the packing and condensation of the DNA to about one tenth of the length without the wrapping structures.

nucleotide: A combination of a nitrogenous base, a pentose sugar, and a phosphate group. The deoxyribonucleotides in DNA are composed of a deoxyribose sugar molecule and a phosphate group, plus one of Guanine, Cytosine, Adenine, or Thymine. The ribonucleotides in RNA are composed of a ribose sugar and a phosphate group, plus one of Guanine, Cytosine, Adenine, or Uracil.

* ochre codon (UAA): One of three stop codons in messenger RNA. The other two mRNA stop codons are UAG (amber) and UGA (opal).

* ochre mutation: Any mutation that changes a DNA nucleotide sequence that normally codes for an amino acid into a sequence that, when transcribed into messenger RNA, becomes the ochre stop codon UAA. See also amber mutation, nonsense mutation, and opal mutation.

* Okazaki fragments: Short, discontinuous fragments of the newly synthesized DNA lagging strand. The growing lagging strand uses as a template the DNA strand which is in the 5' to 3' orientation going from the origin of replication toward the replication fork. Since DNA polymerase can only make a new strand in the 5' to 3' direction from a template that runs in the 3' to 5' direction, it must wait until the double helix unwinds an appreciable amount before it can continue to add fresh nucleotides to the growing lagging strand. This results in the Okazaki fragments, which are later united by DNA ligase to produce an intact DNA lagging strand. They are named after Dr. Okazaki who first described the process.

* **oligonucleotide:** A stretch of about 20 or so nucleotides linked together by phosphodiester bonds.

* **oligonucleotide-directed mutagenesis:** A technique involving the insertion of an oligonucleotide with a specific mutation that is the focus of study into a gene that is then spliced into a cloning vector. It is a means of studying the effect that a gene has on a living system, changing one nucleotide at a time. Also referred to as "site-directed mutagenesis." The technique was pioneered by Dr. Michael Smith of the University of British Columbia. He won the 1993 Nobel Prize in Chemistry for this work.

oncogene: A gene that promotes the onset of cancer. Oncogenes are often mutant forms of normal genes that function in the control of the cell cycle.

one gene – one enzyme hypothesis: A theory put forward in the early 1940s by Dr. George Beadle and Dr. Edward Tatum. In their studies of the red bread mold *Neurospora crassa*, they deduced that a mutation in one gene knocked out the organism's ability to make one enzyme. This concept assumed that every protein is coded for by just one gene. However, it overlooked the possibility that proteins were made up of two or more polypeptide chains, coded for by more than one gene. This rule was later changed to the "one gene – one polypeptide hypothesis."

one gene – one polypeptide hypothesis: This rule states more accurately the relationship between genes and proteins. Each gene codes for one polypeptide chain that can act either as a functional protein by itself or as one chain of several that make up the final 3-dimensional structure of a functional protein.

* **opal codon (UGA):** One of the three stop codons in mRNA. The other two stop codons are UAG (amber) and UAA (ochre).

* **opal mutation:** Any mutation that changes a DNA nucleotide sequence that normally codes for an amino acid into a sequence that, when transcribed into messenger RNA, becomes the opal stop codon UGA. See also nonsense mutation.

* **open reading frame:** The stretch of nucleotides in mRNA that codes for amino acids. It lies between an initiation codon (AUG) and a stop codon (one of UAG, UAA, or UGA).

* **operator gene:** A stretch of DNA that serves as an attachment site for a repressor protein molecule, found only in prokaryotes. When the repressor molecule binds to the DNA it prevents the RNA polymerase enzyme from binding to the promoter region, and hence prevents transcription of the gene in question into mRNA.

operator sequence: See regulatory sequence.

* **operon:** A stretch of DNA that comprises "one complete transcription unit," found only in prokaryotes. An operon is made up of the operator gene, the promoter region, and the structural genes that code for a set of polypeptides. All of the polypeptides produced by one operon play a role in the same process. Two examples of operons found in *E. coli* are the *lac* operon (lactose operon) and the *trp* operon (tryptophan operon). The *lac* operon regulates the production of proteins involved in the uptake and chemical breakdown within the cell of lactose (milk sugar) into its components glucose and galactose. The *trp* operon regulates the production of enzymes required for the synthesis of the amino acid tryptophan.

p arm: See q arm.

* **palindrome:** A word, phrase, or sentence that, when spelled backwards, is unchanged. The term is often applied to unique sequences in DNA in which complementary strands have the same sequence when read in opposite directions. For example, in this double stranded DNA section:

5'- GAATTC - 3'
3'- CTTAAG - 5'

the top strand when read from left to right (i.e., in the 5' to 3' direction) has the same sequence as the bottom strand when it is read from right to left (i.e., also in the 5' to 3' direction). Palindromic sequences in double-stranded DNA serve as binding sites for certain enzymes such as RNA polymerase and restriction endonucleases. It is these palindromic sequences that are the cutting sites for the restriction enzymes. These sequences are usually 4 to 6 base pairs in length, but can be as long as 15 base pairs. (The above example happens to be the recognition sequence for the restriction enzyme called EcoRI). The term is derived from a Greek word which means "to run backwards".

* pBR322: The plasmid, first isolated from *E. coli* cells by two researchers, Dr. Bolivar and Dr. Rodriguez, that proved to be one of the most versatile plasmid cloning vectors. Its entire sequence of 4300 base pairs is known, and it contains many different unique restriction enzyme cutting sites. Because pBR322 also possesses genes for resistance to the antibiotics ampicillin and tetracycline, any bacterial cells that possess the plasmid can be easily selected by plating the bacteria on an agarose plate containing either ampicillin or tetracycline.

peptide: A combination of two or more amino acids linked together by a peptide bond.

peptide bond: The bond that forms between two adjacent amino acids when the amino group of one amino acid reacts with the carboxyl group of another. A water molecule is released in the condensation reaction. All amino acid residues making up protein molecules are linked together by peptide bonds.

peptidyl site (P-site): The site within the ribosomal complex that holds the transfer RNA molecule that is connected to the growing polypeptide chain. (See also aminoacyl site).

peptidyl transferase: The enzyme responsible for catalyzing the formation of a peptide bond between amino acids during the translation of mRNA in the ribosomal complex. It facilitates peptide bond formation between the amino acid that is attached to the transfer RNA molecule that has just docked at the A-site of the ribosomal complex, and the last amino acid that is part of a growing chain attached to the transfer RNA molecule at the P-site of the ribosomal complex. Peptidyl transferase is bound tightly to the larger of the two ribosomal subunits.

phage: See bacteriophage.

phenotype: The observable characteristics of an organism produced by the genes of the organism.

phosphodiester bond: The type of bond that links together the pentose sugar molecules to the phosphate groups along the sugar-phosphate backbone of nucleic acids.

* phosphorus-32 (^{32}P): The radioactive isotope of phosphorus that is much less abundant than the stable isotope, phosphorus-31. It emits streams of beta particles and has a half-life of 14.28 days. In the lab, it is incorporated into biomolecules such as DNA and RNA to track their path and ultimate fate in biochemical reactions.

* physical map: A map of the actual locations of the markers along the DNA molecule, as opposed to a linkage map which is based on the probability of a crossover occurring between the markers. (A physical map and a linkage map of the same stretch of DNA are colinear, but the distances between the markers are not proportional to each other.) The units of measurement used to describe the distances between markers on a physical map are nitrogenous "base pairs." The units of measurement used to describe distances between markers on a genetic linkage map are called "centimorgans" (one centimorgan represents the distance between markers within which there is a probability of 1% that a crossover will occur).

plasmids: Double-stranded, circular molecules of DNA found in certain species of bacteria such as *E. coli*. They exist independently of the bacterial chromosomal DNA and may confer an evolutionary advantage to the host cell (such as resistance to antibiotics). They range in size from 1000 to 200 000 base pairs. The smaller plasmids serve as useful gene-cloning vectors because they have selectable genetic markers on them (such as antibiotic-resistance genes) and several unique restriction enzyme cutting sites that can be used as locations for splicing in pieces of foreign DNA. Also, plasmids can easily be separated from the bacteria's chromosomal DNA and purified.

plasmid cloning vector: A plasmid used as an acceptor of foreign DNA fragments in the formation of recombinant DNA (i.e., gene cloning).

Plasmids that are used as cloning vectors are generally small (around 5000 base pairs long) and replicate in numbers of 10 to 100 copies per bacterial cell (relaxed control). They contain genes that code for proteins that give the host cell resistance to antibiotics such as ampicillin, tetracycline, and kanamycin, and their sequence contains binding sites for restriction enzymes in regions that are not essential for their own replication. (See pBR322.)

point mutation: The substitution of a single nucleotide by another in DNA or, in some viruses, RNA.

* poly-A tail: A polymer of as many as 200 adenosine monophosphate residues that are enzymatically attached to the 3'-end of a freshly transcribed messenger RNA molecule in eukaryotes. This is just a part of the "post-transcriptional modification" that a primary mRNA transcript undergoes before it moves from the nucleus to the cytoplasm. (Note: not all eukaryote primary mRNA transcripts get a poly-A tail. For example, the histone protein mRNAs don't have one).

* polyacrylamide gel electrophoresis: A technique used to separate nucleic acid molecules (based on their different lengths) in a gel matrix. Nucleic acids have a net negative charge on them and will migrate in the direction of a positive charge. At one end of the gel apparatus is a negative electrode and at the other end is a positive electrode. The ends of the gel are placed in contact with a buffer solution and the nucleic acid mixtures are placed in wells at one end of the gel slab. When the current is applied, the nucleic acid molecules begin to migrate through the gel away from the negative electrode toward the positive electrode. Because the shorter nucleic acid fragments are able to squeeze through the gel matrix more easily than the longer ones, the shorter fragments migrate farther along the length of the gel over a given period of time. After several hours, the fragments of DNA or RNA are separated, with the longest fragments near the injection site and the shortest ones at the other end of the gel. Polyacrylamide gels are so fine in their matrix-particle size that they achieve separation of nucleic acid fragments that differ in length by only one nucleotide.

* polycistronic messenger RNA: A large messenger RNA, found only in prokaryotes, that codes for the amino acid sequence of two or more polypeptide chains from two or more cistrons that lie side by side (with each cistron bordered by a start and a stop codon). The proteins coded for by a polycistronic messenger RNA molecule are usually enzymes for the same metabolic pathway.

polymerase chain reaction (PCR): A method used to amplify or copy a specific sequence of DNA in a test tube. The technique is very accurate and quite sensitive. In this technique, one must know the sequence of about 20 nucleotides that immediately flank the target sequence of DNA that you want copied. First, the DNA is heated to a temperature of about 95°C to separate the two complementary strands. "Primers" of single-stranded nucleotide sequences about 20 bases long are added to the test tube along with the four deoxyribonucleoside triphosphates (dGTP, dCTP, dATP, and dTTP). The mixture is then cooled to about 65°C. The base sequences of the "primer" polynucleotides are complementary to the two 20-base sequences that flank the target region of DNA. When the temperature is reduced, they bind to these complementary regions. A DNA polymerase (called Taq polymerase) from a bacterial cell that lives in the heart of hot springs and can function at high temperatures is used to extend the primers in the 5' to 3' direction (as all DNA polymerases do). After several such heating and cooling cycles, you end up with a population of DNA fragments that are exact copies of the sequence that was originally targeted. From just one double-stranded DNA fragment you end up with one million identical fragments after just 20 cycles of the above operation. After 30 cycles, you have one billion copies. The PCR technique (which has often been referred to as a "DNA photocopier") was pioneered by Dr. Kary Mullis who won the Nobel Prize in 1993 for his discovery.

polygenic: A physical characteristic that is manifested due to the influence of more than one pair of alleles.

polynucleotide: A chain of nucleotides linked together by 3'-5' phosphodiester linkages.

polypeptide: A chain of amino acids linked together by peptide bonds. By itself, a polypeptide chain can be an entire protein molecule, or it can be one of two or more polypeptide chains that constitute the final structure of the functional protein (e.g., the hemoglobin molecule is composed of four separate polypeptide chains).

* post-transcriptional modifications: Chemical modifications made to the messenger RNA molecule in eukaryotes immediately after it has been transcribed but before it leaves the nucleus to enter the cytoplasm for translation. Examples of these are the addition of a methylated guanosine cap to the 5' end of the precursor mRNA, the addition of adenine residues to its 3' tail, and the removal of introns and splicing together of the exons by the spliceosome to produce the mature mRNA molecule.

* post-translational modifications: Chemical modifications made to polypeptide chains after they have been freshly synthesized. Examples of these include the removal of one or several amino acids from within or from either end of the polypeptide chain; the chemical modification of some amino acid R groups such as addition of a hydroxyl group; and the attachment of sugar groups to certain amino acids in some proteins.

precursor mRNA: The freshly transcribed molecule of messenger RNA before any post-transcriptional modification has taken place. Also called primary transcript or pre-mRNA.

Pribnow box: A sequence of six nucleotides (i.e., TATAAT) on the coding strand of prokaryotic DNA (named after Dr. David Pribnow) that serves as a binding site for RNA polymerase for transcription of mRNA. It is about 10 base pairs upstream (i.e., on the 5' side) of the transcription start site. This sequence of nucleotides in prokaryotes is a part of the overall promoter region which spans about 60 base pairs in length upstream from the transcription start site. In addition to the TATAAT sequence, another sequence of 6 nucleotides (TTGACA) which is 35 base pairs

upstream from the transcription start site also contributes to the specific binding of RNA polymerase in *E. coli* cells.

Pribnow Box

5'- - -TTGACA - - - - - -TATAAT- - - - - - Transcription Start Site

-35 base pairs   -10 base pairs   0 point

The Pribnow Box plays a role in the initiation of transcription in prokaryotes, while the Hogness Box does the same job in eukaryotic DNA. (See also Hogness Box.)

primary structure: The linear order of amino acids in a polypeptide chain.

primary transcript: See precursor mRNA.

* primase: The enzyme responsible for catalyzing the synthesis of the short stretch of primer RNA (about 10 ribonucleotides long) which is necessary for DNA replication.

probe: Usually a stretch of single-stranded DNA, radioactively labeled with Phosphorus-32, used in the lab to seek out and bind to a complementary sequence of nucleotides on a target DNA molecule that is part of a large population of DNA molecules. Such a radioactive DNA probe is often used to mark a specific gene of interest on a genomic DNA sample that has been cut with restriction enzymes and then separated by electrophoresis. Because the probe is radioactively labeled, it will create a dark signal spot when exposed on an X-ray film, thereby marking the location of the gene of interest.

prokaryctic cell: A cell that lacks a membrane-bound nucleus (thus, its DNA is not contained within a nucleus) and a cytoskeleton; the single-celled organisms, bacteria, and cyanobacteria are prokaryotes.

promoter: The region of DNA that "promotes" transcription and is upstream (i.e., on the 5' side) from the transcription start site. It serves as the binding site for RNA polymerase, which transcribes the DNA template strand into a complementary RNA molecule. (It could be a messenger RNA, ribosomal RNA, or transfer RNA molecule that is transcribed.) The rate of RNA transcription depends on the sequence of these nucleotides (i.e., some promoters are stronger than others, having a greater affinity for the binding of an RNA polymerase molecule and resulting in a higher rate of transcription).

promoter sequence: See regulatory sequence.

protease: Any enzyme that digests proteins by hydrolysis.

protein: A molecule that is composed of one or more polypeptide chains. Each polypeptide chain is in turn composed of a linear sequence of amino acids linked together by peptide bonds.

* protein kinase: A class of enzymes that transfer a phosphate group from ATP to one of three specific amino acid residues of certain proteins.

proteolytic enzymes: Any enzyme that digests proteins into smaller polypeptide chains (see protease).

* pseudogene: A gene that, as its name implies, is a "false" gene. It has a nucleotide sequence that is similar (but not identical) to a known functional gene that resides at a different locus on the chromosome. A pseudogene is non-functional because nucleotide substitutions, deletions, or additions occurred through time, preventing normal transcription and subsequent translation.

P-site: See peptidyl site.

* pulsed-field gel electrophoresis: A form of agarose gel electrophoresis that separates DNA fragments of different lengths by using alternating pulses of electric current in two directions that are at 90° to each other. This has the effect of "straining" larger DNA fragments through a gel matrix. One can separate very large fragments of DNA with this technique. (Whole yeast cell chromosomes, with a length of 1 800 000 base pairs, have been separated from smaller fragments with this technique).

purines: A class of nitrogen-containing organic bases, with a double-ring structure, in which the nitrogen atoms are always at specific locations in the rings. The purines that occur in both DNA and RNA are Adenine and Guanine.

pyrimidines: A class of nitrogen-containing organic bases with a single-ring structure, in which the nitrogen atoms are always at specific locations in the ring. The pyrimidines that occur in DNA are Cytosine and Thymine, whereas the pyrimidines that occur in RNA are Cytosine and Uracil.

q arm: Each eukaryotic chromosome has a centromere located somewhere along its length. The exact location of the centromere differs between chromosomes. The pinched-in region of the centromere results in the chromosome having two "arms." The longer arm of a chromosome is called "q arm" and the short arm is the "p arm."

quaternary structure: The final 3-dimensional structure of a protein molecule that is made up of two or more polypeptide chains.

* reading frame: A sequence of nucleotides in either DNA or mRNA that begins with a start codon, has a middle section of triplet codons that code for amino acids, and ends with a stop codon.

recessive allele: An allele that is only expressed in a diploid organism when it exists in the homozygous state. In the heterozygous state, the dominant allele masks the presence of the recessive allele. In some cases, the dominant allele of a pair of alleles codes for a normally functioning protein molecule and recessive alleles produce a protein that does not function properly, resulting in a genetic disorder such as sickle-cell anemia or Tay Sachs disease. In some cases, the defective protein has no ill effects on the individual. For example, the allele for blue eye color prevents the formation of eye pigment, leaving the eyes blue.

recombinant DNA (chimeric-DNA): In genetic engineering terms, a DNA molecule that is constructed by the *in-vitro* splicing together of pieces of DNA from different sources.

recombinant DNA technology: A branch of molecular biology that involves the splicing together of DNA from different organisms and transferring these composite DNA molecules into the cells of bacteria and other living organisms. The desired results of recombinant DNA technology are 1) to isolate specific pieces of DNA from any source for the purpose of multiplication and subsequent purification for further analysis; 2) to isolate specific genes that produce a protein product that is useful in the fields of agriculture, industry, or medicine (e.g., insulin); and 3) to develop new characteristics in organisms, such as herbicide-resistant crop plants and fruit that remains fresh longer.

* regulatory gene: In prokaryotes, a gene that codes for a protein molecule called a "repressor" which acts to regulate the expression of structural genes. The repressor molecule binds to a specific operator gene thus preventing transcription of the structural genes in that particular operon.

* regulatory sequence: A DNA sequence that does not code for a protein but functions by regulating the expression of a gene. (Also called a promoter sequence in all organisms; operator sequences are found in prokaryotes.) A promoter sequence of nucleotides serves as a binding site for RNA polymerase for subsequent transcription of a gene. An operator sequence of nucleotides serves as a binding site for a repressor protein which halts transcription of a gene.

* reiterated genes: Groups of copies of the same gene that are clustered together on a chromosome. These multi-gene families can be observed for ribosomal RNA genes and transfer RNA genes as well as the genes that code for the histone proteins.

* release factors: Special proteins found in the cell cytoplasm that mediate the release of the finished polypeptide chain from the ribosomal complex upon completion of translation of the mRNA.

replication fork: DNA replication involves the separation of the two strands of the double helix. Where the two strands separate, the junction between the unwound double helix and the two single strands forms a Y shape. This location is called the replication fork. The two exposed single strands can then act as templates for the synthesis of two complementary strands of DNA from free-floating deoxyribonucleoside triphosphates.

* replication origin: A sequence of nucleotides on the DNA at which DNA synthesis begins. In prokaryotes, there is only one replication origin on the circular DNA molecule. In eukaryotic DNA, there are many replication origins.

* replisome: The collective term given to DNA polymerase plus the other enzymes that act in unison at the replication fork of bacterial DNA to carry out DNA synthesis.

* repressible system: A bacterial system consisting of a regulatory gene and its operon. Transcription of a structural gene is inhibited or "repressed" when the product of the enzyme it codes for is present in ample amounts in the cell. One of the best-understood repressible systems is the tryptophan operon system of *E. coli* cells. If a lot of the amino acid tryptophan is floating around in the cell, the genes that code for the enzymes in the metabolic pathway for the synthesis of tryptophan are shut down. This saves energy and materials. The shutdown occurs as follows: a regulator gene codes for a protein called an aporepressor which binds to a freely floating tryptophan molecule which acts as a co-repressor. The aporepressor/co-repressor complex binds to the operator gene of the tryptophan operon. Because this blocks off the promoter region, the RNA polymerase molecule cannot bind to it to start transcription. (See repressor protein.)

* repressor protein: A protein molecule coded for by a "regulator" gene. Repressor protein molecules may act alone or as "aporepressors" that require a co-repressor (such as an amino acid). In either case, the active repressor binds to the operator gene of a specific operon and prevents transcription of the structural genes. The operator region is usually downstream (i.e., to the 3' side) of the promoter region so the bound aporepressor/co-repressor complex blocks the binding and/or the advance of RNA polymerase. The result: no transcription can occur.

restriction endonuclease (restriction enzyme): An enzyme that cleaves double-stranded DNA at special sites within DNA called restriction sites or recognition sequences. In most cases, the result of a single cut is two fragments with protruding 5' ends called "sticky ends." These protruding 5' termini are single strands of DNA that are complementary to each other. In the example below, the palindromic cutting site of the restriction enzyme called EcoRI is used:

$$5' - - - G - A - A - T - T - C - - - 3'$$
$$3' - - - C - T - T - A - A - G - - - 5'$$

The cutting of the DNA at the sites indicated by the arrows yields two fragments of DNA, each with protruding 5' ends. These are called "staggered ends" or "sticky ends" because they can be re-attached to each other by complementary base pairing.

$$5' - - - G \qquad\qquad 5'A - A - T - T - C - - -3'$$
$$3' - - - C - T - T - A - A \, 5' \qquad G - - -5'$$

Because a restriction enzyme will cut DNA from any source at exactly the same site and will leave exactly the same kinds of complementary single-stranded ends, you can splice DNA from two completely different organisms. This splicing together of DNA from different species is the basis of recombinant DNA technology. More than 800 different restriction endonuclease enzymes have been discovered in different bacterial species that recognize and make cuts at about 100 different palindromic DNA sequences. The name of a restriction enzyme is an abbreviation, specifying the species and strain of the bacteria it comes from plus a Roman numeral showing the chronological order of discovery of that particular protein enzyme in that strain of bacteria. For example, the term "EcoRI" means that the restriction enzyme comes from *Escherichia coli* bacteria, strain-type R and that it was the first one to be isolated from *E. coli* "R" strain cells. Although most restriction enzymes cut DNA to yield "staggered" complementary ends, some cut DNA to leave blunt ends instead. Restriction enzymes are believed to exist within bacterial cells to protect against the invasion of viral DNA. If a piece of foreign DNA with restriction enzyme cutting sites on it enters the cell, the enzymes will cut it up into little pieces, rendering it harmless.

restriction fragment: A piece of double-stranded DNA which was generated by cutting a longer length of DNA with a restriction endonuclease.

* restriction fragment length polymorphisms (RFLPs): The different lengths of DNA fragments generated by cutting the DNA from different individuals with the same restriction endonuclease enzyme. The cutting sites of any restriction enzyme are the "palindromic sequences" that are scattered along the length of chromosomal DNA. The span of DNA that runs between any two sites that can be cut by one particular restriction enzyme can vary greatly between different individuals of the same species. Therefore, the lengths of the DNA fragments obtained by cutting the DNA of two different people with the same restriction enzyme would also be different. These variations are produced by mutations that can either create a palindromic cutting-site or make a pre-existing one disappear. Another phenomenon which contributes greatly to fragment length differences is the presence of repetitive nucleotide sequences. These sequences range from two to as many as 100 base pairs in length, according to some reports. The sequences may be repeated from about 10 to over 500 times. Because the number of repeats varies greatly in a population, the term "variable number of tandem repeats," or VNTRs, is applied to the regions. A simple example of a VNTR using a dinucleotide repeat of Adenine and Cytosine repeated over and over again on the same strand of DNA would be ACACACACACACAC. VNTRs between any two restriction enzyme cutting sites will, of course, change the lengths of DNA fragments in that individual. For example, if person A has more VNTRs between two cutting sites than person B has, than person A's restriction fragment lengths would be longer than person B's. RFLPs are used frequently in forensic analysis where the RFLP pattern of DNA taken from a tissue sample left at the scene of a crime (i.e., blood, semen, hair, or skin) is compared with the RFLP pattern of the DNA taken from a suspect. If the two RFLP patterns match (the patterns resemble a barcode that you see on grocery store items), then there is a strong probability that the tissue sample left at the scene of the crime belongs to the suspect. (See DNA fingerprint technique).

restriction site: A sequence of nucleotides along the DNA double helix at which a particular restriction endonuclease cuts the molecule, leaving either staggered or blunt ends.

retroviruses: A class of viruses that contain RNA as their principle genetic material. Retroviruses violate the old "central dogma" of genetics which stated that the flow of genetic information is in one direction only (from DNA to RNA to protein). These viruses contain the information necessary for the production of an enzyme called reverse transcriptase which uses the viral RNA molecule as a template to make a complementary DNA strand. The HIV particle which is believed to cause AIDS is a retrovirus.

reverse transcriptase: The enzyme used by retroviruses to make a complementary single strand of DNA using an RNA molecule as a template. The enzyme then uses the newly-formed single-stranded DNA molecule as a template for the synthesis of the complementary sister strand of DNA to make a complete double helix.

ribonuclease: An enzyme that digests RNA molecules by hydrolysis.

ribonucleic acid: A polymer of nucleotides that have ribose (instead of deoxyribose) as their pentose sugar component plus one of the four nitrogenous bases, Guanine, Cytosine, Adenine, or Uracil (instead of Thymine as in DNA). The three classes of RNA are messenger RNA, ribosomal RNA and transfer RNA.

ribonucleoprotein: A combination of nucleic acid and protein, used in reference to ribosomes and snRNPs. (See also small nuclear ribonucleoproteins.)

ribonucleotide: An organic molecule made of a ribose sugar molecule and a phosphate group, plus one of Guanine, Adenine, Cytosine, and Uracil.

ribosomes: Cytoplasmic organelles that are the site of translation of mRNA (protein synthesis). Ribosomes are made up of two different-sized subunits, each of which is made up of almost equal parts of protein and ribosomal RNA. Ribosomes are roughly 10-20 nanometers in diameter.

* ribozymes: RNA molecules that act as enzymes.

Ribozymes were first discovered in 1982 when Dr. Thomas Cech and his team looked at the splicing mechanism of an intron in the ribosomal RNA of a ciliated protozoan called *Tetrahymena thermophilia*. They showed that no protein enzyme was involved in splicing out the intron, and that the excision of the intron with subsequent resplicing was mediated by the RNA molecule itself. Prior to this Nobel Prize-winning discovery, it was believed that all enzymes were proteins.

RNA polymerase: An enzyme that catalyzes the transcription of an RNA molecule (either messenger RNA, ribosomal RNA, or transfer RNA) from a DNA template strand. In prokaryotes, there are two kinds of RNA polymerase. One, called a "primase," produces an RNA primer strand of about 10 ribonucleotides which acts as a primer for DNA replication. The other kind transcribes all three other kinds of RNA. In eukaryotes, each of the three kinds of RNA is transcribed by a different RNA polymerase molecule.

* satellite DNA: Highly repetitive, short sequences (5 to 10 base pairs) of DNA, usually found at the centromere region of eukaryotic chromosomes. Satellite DNA gets its name from a laboratory observation: when eukaryotic DNA is isolated from a cell and centrifuged at high speeds in a density gradient, usually three bands are visible in the centrifugation tube. The thick, middle band, comprising the majority of the cellular DNA, is sandwiched between two thinner bands of DNA which have slightly different base compositions. The upper and lower thinner bands hover like "satellites" above and below the main band. The upper band of satellite DNA is composed of highly repetitive, A-T-rich sequences, making it less dense so it floats on top. The lower band of satellite DNA is composed of highly repetitive, G-C-rich DNA, making it dense so it sinks to the bottom of the density gradient. Satellite DNA is composed of non-coding sequences that are repeated with a very high frequency and occur at or near the centromere and the telomeres of a chromosome. Near the centromere, they may serve as a binding site for proteins that attach to the microtubules of the spindle apparatus. Satellite DNA that occurs at the telomeres (the ends of a linear eukaryotic chromosome) is believed to provide structural stability to the chromosome.

semi-conservative replication: In DNA replication, the idea that each of the two new molecules has within its double helix an intact single strand from the original DNA molecule. Early scientists thought that DNA replicated by "conservative replication," in which the two original strands rejoined and the two new strands hydrogen bonded to each other. In 1958, Dr. Matthew Meselson and Dr. Franklin Stahl of the California Institute of Technology demonstrated the phenomenon of semi-conservative replication of DNA by labelling the DNA of *E. coli* with ^{15}N and observing the location and concentration of the ^{15}N after each round of replication.

* sequence tagged site (STS): A short stretch of DNA, usually about 200 to 500 base pairs long, with a unique sequence and a known location within the genome. Because the job of sequencing the entire Human Genome is a multinational task involving many labs in many different countries, the sequencing data is often obtained by different strategies of mapping and sequencing. In order to collate all of this DNA sequencing data into a universally understandable format, sequence tagged sites offer quick and easy reference points or "map signposts." For example, if a scientist wants to continue to sequence a stretch of DNA where another lab left off, all that he or she needs to do is to "fish out" the STS that is known to be near that particular DNA region of interest. In the lab, STSs can be radioactively labeled and used as probes to isolate cloned DNA fragments that may contain genes of interest.

sex chromosomes: A pair of chromosomes that determine the gender of the organism.

sex linkage: Any gene found only on the X chromosome or only on the Y chromosome. Previously, the term sex-linkage was used to denote genes on the X chromosome only. However, a few (very few) genes on the Y chromosome have been identified. Genes on the X chromosome are now called X-linked and genes on the Y chromosome are called Y-linked. In some species, individuals with two X chromosomes are female and those with one X and one Y chromosome are male. Since males have only one X chromosome, they have only one allele for each of the X-linked genes. Therefore, they express any recessive alleles that may be located on the X-chromosome. For example, a gene for color vision is X-linked. Only 1% of females are colorblind while 8% of males are. Some genes with deleterious recessive alleles are X-linked. This is why diseases such as hemophilia and Duchenne muscular dystrophy nearly always occur only in males.

* shuttle vector: A molecule of plasmid DNA that has been genetically engineered in such a way that it can replicate in both prokaryotic and eukaryotic cells. The plasmid-vector DNA molecule can be spliced to a fragment of foreign DNA under investigation, and this recombinant DNA molecule can then be used to "shuttle" the inserted genes from one type of cell to another (in particular from a prokaryotic cell to a eukaryotic cell).

sickle-cell anemia: A genetic disorder that occurs in individuals that are homozygous-recessive for that genetic trait. This means that they carry two recessive alleles for sickle-cell anemia within their chromosomes. The genetic defect results in the placement of the amino acid valine in position #6 of the beta chain of hemoglobin instead of the normal amino acid, glutamic acid. Because of this replacement, the shape of the hemoglobin molecules are distorted when they become deoxygenated. The hemoglobin molecules stack into thin-sided crystals within the red blood cell, changing the surface tension of the cell. Consequently, the red blood cells lose their characteristic disc-shape and form elongated sickle-shapes which are fragile and burst easily. Small-diameter blood vessels such as capillaries can be clogged by sickled cells, resulting in oxygen starvation of adjacent cells. Physical symptoms of sickle-cell anemia are general weakness, heart failure, bone infections, extreme pain in the joints, kidney failure, and early death of the affected individual.

* sigma factor: A small polypeptide subunit

found within RNA polymerase of the bacterial cell *E. coli*. It recognizes specific binding sites on DNA molecules for the attachment of RNA polymerase to start transcription.

* **silent mutation:** A point mutation that changes a codon but does not change the amino acid that is incorporated into the protein. For example, if the third nucleotide of the nucleotide triplet GGA changed to a T to give GGT, the amino acid glycine would still be coded for.

* **single-stranded DNA binding protein:** A high-molecular-weight protein molecule that binds to the single strands of DNA after the double helix is unzipped by the action of the enzyme helicase during DNA replication in *E. coli*. The function of the SSB protein is to stabilize the single strands and prevent them from re-forming hydrogen bonds with each other or within the single strands themselves.

* **site-directed mutagenesis:** See oligonucleotide-directed mutagenesis.

* **small nuclear ribonucleoproteins (snRNP):** Small particles located in the nucleus of eukaryotes that are composed of RNA and protein molecules. Some of these snRNPs form parts of spliceosomes. The snRNP molecules are typically about 150 nucleotides long and are associated with about seven or more protein molecules within the spliceosome. (See spliceosome).

* **Southern blotting technique:** A method of visualizing DNA fragments separated by gel electrophoresis to make a DNA fingerprint or for other applications. After the DNA fragments have been separated according to size with agarose gel electrophoresis, the DNA is treated with sodium hydroxide to "denature" it (separate the complementary strands). The bands of DNA fragments are then transferred to a nitrocellulose sheet by drawing a buffer through the gel and the sheet. The nitrocellulose is exposed to a radioactive probe which binds to complementary sequences. Excess probe is washed away and the nitrocellulose sheet is placed on a piece of X-ray film. Darkened bands on the film reveal the locations of the DNA fragments that were complementary to the probe.

* **spliceosome:** A tiny organelle within the nucleus of eukaryotic organisms that splices out the introns that occur within the precursor, mRNA. The spliceosome is a complex of small nuclear ribonucleoprotein molecules (snRNPs) which are small RNA molecules bound to about seven different proteins. The freshly transcribed RNA molecule combines with a spliceosome. Then, the spliceosome interacts with the few nucleotides at each end of an intron and cuts the RNA to release the intron. It then splices together the two exons that were separated by that intron.

split genes: Genes in eukaryotic organisms in which the coding regions (exons) are separated by non-coding regions (introns). Split genes occur in eukaryotes and in some animal virus genomes but are not found in prokaryotes.

staggered cuts: The type of cutting action that certain restriction enzymes make on double-stranded DNA molecules. The resulting ends of the DNA at the cut-site are staggered: they have overhanging single-stranded 5' ends (sticky ends). (See restriction endonuclease).

start codon: A nucleotide triplet that codes for the first amino acid of a protein chain. The start codon in mRNA is AUG. The start codon codes for the amino acid N-formyl methionine in prokaryotes and plain methionine in eukaryotes. (Also called the initiation codon.)

sticky ends: The complementary single-stranded ends of DNA that result from the cutting action of a restriction enzyme. (see staggered cuts).

structural gene: A sequence of nucleotides in DNA that codes for a sequence of amino acids in a polypeptide.

substitution mutation: A mutation that involves the substitution of one nucleotide by another.

sulfur-35 (^{35}S): The radioactive isotope of sulfur that is much less abundant than the stable isotope, sulfur-32. Sulfur-35 emits streams of beta particles and has a half-life of 87.2 days. In the lab, it is incorporated into biomolecules such as proteins to track their path and ultimate fate in biochemical reactions.

* synthetic linkers: Short segments of double-stranded DNA molecules that are chemically synthesized *in vitro* and then spliced into a cloning vector such as a plasmid molecule. These synthetic linkers are custom-made, and usually contain one or more palindromic sequences that can be recognized and cut by certain restriction endonucleases. This is a handy gene-splicing technique because the genetic engineer can take a vector molecule that has a synthetic linker spliced into it and use a restriction enzyme that matches the site in the linker to cut it open. He can then splice in a fragment of foreign DNA that has been generated by cutting with the same restriction enzyme. This means that both the open ends of the vector molecule and the terminal ends of the fragment of interest will have complementary "sticky ends" and can bind to each other. With the addition of DNA ligase to seal the sugar-phosphate backbone at the splice site, you end up with a new recombinant DNA molecule (i.e., vector DNA, synthetic linker DNA, and DNA fragment of interest all spliced together).

* tandem repeat: See variable number tandem repeat sequence.

* Taq DNA polymerase: An enzyme that catalyzes the synthesis of DNA within a bacterial cell called *Thermus aquaticus*. What is unusual about this bacterial cell is that it lives in the heart of hot springs where the temperature nearly reaches the boiling point of water. The Taq DNA polymerase enzyme is heat stable and can function quite well after exposure to temperatures of 95°C (a temperature that would denature and hence destroy most other protein enzymes). For this reason it is useful in the polymerase chain reaction technique (PCR) which is used to clone fragments of DNA in a test tube. One of the steps in the PCR technique requires a reaction temperature of close to 95°C to separate the two strands of a DNA double helix. (See polymerase chain reaction.)

* TATA box: See Hogness Box.

* tautomeric shift: The reversible shifting of a hydrogen atom within a molecule to a new position within the same molecule. The result is that the molecule is converted from one type of isomer to another. When a tautomeric shift occurs within a nitrogenous base of a DNA molecule, it can change the base-pairing specificity of the nitrogenous base that underwent the shift. Thus, it can cause a point mutation.

* telomeres: The ends of linear eukaryotic chromosomes. They show structural similarity between different organisms. The similarity is in the nucleotide sequences of the short DNA segments, with one strand being rich in Guanines and the complementary strand, of course, being rich in Cytosines that are tandemly repeated. It is believed that the telomeres provide structural stability to the chromosome. The tandemly repeated sequences in telomeres are non-coding and are added to the chromosome ends by an enzyme called telomerase.

template strand: The strand of DNA that is transcribed into a complementary messenger RNA molecule. The nucleotide sequence of the template strand is complementary to the sequence in its partner strand, the coding strand, in the DNA double helix. The template strand is sometimes referred to as the "anti-sense" strand.

thymine: One of the four nitrogenous bases found in DNA.

transcription: The formation of an RNA molecule (either a messenger RNA, a transfer RNA, or a ribosomal RNA) from a complementary single-stranded DNA template by the action of the enzyme RNA polymerase.

transcription factors: DNA binding proteins that help the RNA polymerase bind with the promoter and initiate RNA synthesis.

transfer RNA: An RNA molecule that attaches to a specific amino acid in the cytoplasm and transfers it to a site within the ribosomal complex. While held in place by the tRNA, the amino acid is added to the growing polypeptide chain during translation. Transfer RNA molecules are usually about 75 nucleotides long, and their structure is characteristically a "clover-leaf" configuration when flattened out in two dimensions.

transgenic animals: Animals whose genetic constitution has been modified by genetic engineering techniques. How is a transgenic animal produced? First, isolate a fertilized egg of the organism (such as the zygote of a sheep). Then, using a microscopic syringe, inject foreign DNA containing the genes of interest into the zygote. A few of the sheep zygotes that are injected with the foreign DNA will combine the injected DNA into one of their own chromosomes. Remember that each and every cell produced from that fertilized egg cell will contain the recombinant DNA in its chromosomes as well. This includes the germ-line cells (sex cells). Reasons for making transgenic animals include: (1) having an animal such as a sheep produce proteins that normally are never made in that species (an example is the human blood-clotting factor called Factor IX which can be made to be secreted in sheeps' milk); (2) having animals that grow leaner and faster; and (3) possibly engineering an animal with human histocompatibility proteins so the animal's organs could be used for transplants.

translation: The process of polypeptide chain formation (protein synthesis) that is mediated by the nucleotide sequence of a messenger RNA molecule. Translation begins when the messenger RNA molecule forms a complex with the ribosome. Subsequently, the codons of the mRNA are read and translated into amino acids by amino acid-carrying transfer RNA molecules.

* transposons: (transposable elements or "jumping genes") Mobile segments of DNA that can move from one location on a chromosome to another or to a different chromosome. In bacterial cells, a transposon can move from one location of the chromosome to another, from a plasmid to the chromosome, or from the chromosome to a plasmid. Throughout evolution, this jumping around has resulted in the transposons picking up and retaining little bits of genetic material here and there that proved to be beneficial to the survival of the bacterial cells that harbor them. For example, over the course of time transposons have picked up several genes for resistance to antibiotics from several different plasmids and combined them into a single plasmid. Transposons are also important

components of eukaryotic genomes. They were first reported by Dr. Barbara McClintock at Cold Spring Harbor in the 1940s. She studied the color changes that take place in kernels of Indian corn and postulated that these changes occurred because mobile genetic elements moved from one position in the genome to the genes that control kernel color. She was awarded the Nobel Prize in 1983 for being the first to identify transposons.

tritium (hydrogen-3 or ^{3}H): The radioactive isotope of hydrogen that is much less abundant than the stable isotope, hydrogen-1. Tritium emits weak rays of beta particles and has a half-life of 12.32 years. In the lab, tritium is incorporated into organic molecules used in experiments to trace the path and fate of the molecule.

tumor: A mass of cells caused by abnormal and uncontrolled cell division (mitosis). There are two classes of tumors: benign tumors (such as warts) which are not life-threatening, and malignant tumors (such as melanoma) which can break away from their site of origin and spread throughout the body. Most malignant tumors, if left untreated, can be fatal.

universality of the genetic code: With very few exceptions, every living organism on the planet uses the same genetic code of 64 codons (each codon being a set of three nucleotides) to code for the 20 amino acids as well as the "start" and "stop" codons in protein synthesis. The few exceptions to this rule include the genetic code of mitochondrial DNA in mammals and yeast cells, chloroplast DNA, the DNA of some ciliated protozoans, and the DNA of a mycoplasm. Below is a list of these few differences in the genetic code.

Yeast mitochondria	CUA codes for Threonine instead of Leucine; UGA codes for Tryptophan instead of STOP.
Mammalian mitochondria	AUA codes for Methionine instead of Leucine; AGA and AGG code for STOP instead of Arginine; UGA codes for Tryptophan instead of STOP.

| *Mycoplasma capricolum* | UAG codes for Tryptophan instead STOP. |
| Four ciliated protozoans: *(Tetrahymena thermophilia, Stylonychia lamnae, Paramecium tetraurelia,* and *Paramecium primaurelia.)* | UAA and UAG code for Glutamine instead of STOP. |

uracil: One of the four nitrogenous bases found in RNA.

* variable number tandem repeat sequences (VNTRs): A non-coding nucleotide sequence from two to 100 nucleotides long that are repeated side-by-side (hence the term "tandem") and scattered throughout the genome. For example, the tandem repeat of the dinucleotide sequence of AC, giving the pattern of ACACACAC etc., occurs frequently in the human genome. Another example of a tandem repeat is the trinucleotide sequence of CAT in a non-coding region of DNA. This tandem repeat would look like this: CATCATCATCAT etc. Each person has a different number of these tandem repeats within his or her genome. This fact makes the science of restriction fragment length polymorphism useful in DNA fingerprinting. A core sequence of 10 to 15 base pairs is often found in VNTRs. Thus, a probe that is complementary in nucleotide sequence to this core sequence can be used to reveal the lengths of restriction fragments containing these VNTRs. This reveals the DNA fingerprint of an individual. (See also restriction fragment length polymorphism).

* viroid: A group of plant pathogens that consist solely of naked RNA which is usually about 270 to 380 nucleotides long. They affect plants by stunting their growth and derailing the regulatory systems that control their normal development. The nucleotide sequences of viroids are very similar to the nucleotide sequences of a certain group of self-splicing introns found in the ribosomal RNA molecules of eukaryotes. This suggests that the viroids originated as introns that escaped from eukaryotic genes (introns are capable of excising themselves from the rest of the gene without the help of any other enzyme). Viroids are thought to act as "anti-sense" strands that are complementary to (and hydrogen bond with) messenger RNA molecules that are normally transcribed in the plant cells which they invade. If this is so, then the cell is unable to transcribe the duplex RNA molecules that are formed.

virus: A tiny infectious particle, much smaller than a typical cell, composed of an outer protein coat containing a nucleic acid. The protein coat of different viruses have different shapes. Inside the protein coat, the nucleic acid can be either DNA or RNA. A virus can replicate itself only after it enters a host cell where it takes over the metabolic machinery of the cell to make copies of itself. Outside a living cell, it cannot replicate itself and remains inanimate. Some RNA-containing viruses, called retroviruses, make a DNA molecule from the virus' RNA template inside the host cell. Viruses lie on the border-line of classification as living or non-living, since they remain inanimate outside living cells and only exhibit the characteristics of living things after they enter living cells.

Watson-Crick model of DNA: The now-famous double-helix model of DNA that was elucidated in 1953 by James D. Watson and Francis H. C. Crick. It exists most often in living cells in the hydrated "B-form" (as opposed to either the dehydrated "A-form" or the "Z-form"). It is a "right-handed" double helix, which means that if you could look down the long axis of the molecule, you would see that the two chains wind around each other in a clockwise (or right-handed) direction, with 10 nucleotides per 360° turn. The plane surfaces of the base pairs are perpendicular to the long axis of the double helix. The outer sugar-phosphate backbone of the double helix helps to maintain the structural integrity of the centrally positioned complementary base pairs that make up the genetic code.

* wobble hypothesis: An hypothesis used to explain why the anti-codon of some transfer RNA molecules can recognize more than one different codon on the messenger RNA molecule during translation into an amino acid chain. If there were a tRNA molecule with a unique anti-codon for each of the 61 amino acid codons of messenger RNA, then there would have to be 61 different tRNA molecules. In fact there are only about 45 different types of transfer RNA molecules to accommodate all 61 messenger RNA codons that code for the 20 amino acids. (The other three mRNA codons code for termination.) These 45 tRNA molecules are enough because of the "wobble effect." Some tRNA anti-codons can recognize two or more mRNA codons. For example, the base U (Uracil) in the third position of a tRNA anti-codon can pair with either an A or G in the third position of an mRNA codon. This relaxation of the complementary base-pairing rule is called "wobble." Even greater flexibility exists within tRNA anti-codons that contain the modified nitrogenous base called Inosine. (Inosine is made by the enzymatic modification of Adenine after tRNA is synthesized). When Inosine is present in the third position of a tRNA anti-codon, it can base-pair with either U, C, or A in the third position of the mRNA codon. For example, a tRNA anti-codon that reads GGI (the I stands for Inosine) can bind with the complementary mRNA codons CCU, CCC, or CCA, all of which code for the amino acid proline.

X-chromosome: The "female" sex chromosome in mammals and several other species. When two X chromosomes occur in the diploid state of the organism, the organism is female. When only one X-chromosome occurs along with a Y chromosome in the diploid state of an organism, then the organism is male.

X-linkage: All genes on the X chromosome are said to be "X-linked" and thus form a linkage group.

XXY trisomy: (Klinefelter's Syndrome) A chromosomal abnormality in humans that results in a person with two X chromosomes and a single Y chromosome. This condition results in a sterile male with small testes and little body hair. Many exhibit some female sexual characteristics such as breast development. During meiosis of a diploid female primordial sex cell, the X chromosomes fail to segregate during either anaphase I or anaphase II. This results in a haploid egg cell with two Xs. When the egg with two Xs fuses with a sperm cell with one Y chromosome, the resulting zygote has an XXY chromosomal content.

XYY trisomy: A condition in which the individual's cells carry one X and two Y chromosomes. XYY males are frequently taller than the average male and may have a mildly reduced mental function. XYY males are usually fertile. The XYY condition results from the failure of the Y chromosomes to segregate during either anaphase I or anaphase II of meiosis in the spermatogonia, producing YY sperm.

Y chromosome: The chromosome found only in the male sex of all mammals and many other species. (But, for example, not in birds.) In humans, the Y chromosome is much smaller than the X chromosome. Very few genes on the Y chromosome have been identified. However, one gene on the Y chromosome is known to trigger male development in the embryo.

yeast: Single-celled eukaryotic organisms that belong to the fungus family and reproduce by budding. The most well-studied and useful of the yeasts is *Saccharomyces cerevisiae*, which is used to make bread and beer. Yeast is also very useful in gene cloning experiments in which the enzymes of a eukaryotic cell are necessary to process and correctly modify the product of a cloned gene.

* yeast artificial chromosome (YAC): An artificially synthesized yeast chromosome made for the purpose of cloning very large fragments of foreign DNA up to a half a million base pairs long. YACs contain two telomeres and a centromere, so when it is inserted into a yeast cell, it divides and migrates to opposite poles of the cell during mitosis just as one of the yeast's own chromosomes would. YACs were developed in 1987 by Dr. David T. Burke, Dr. George F. Carle, and Dr. Maynard V. Olson at the Washington University School of Medicine. Several human DNA sequences of interest are very large. For example, the gene that codes for the human blood-clotting factor VIII spans about 190 kilobases within the human X chromosome. In order to clone such a large DNA segment, a YAC vector must be used.

Y fork: The Y-shaped region in the part of a DNA molecule that is undergoing replication. The two template strands of the original double helix form the arms of the capital letter Y and the unwound double-stranded DNA forms the stalk of the letter Y.

* Z-DNA: A conformational form of DNA that was discovered in 1973 by Dr. Alexander Rich and his colleagues at the Massachusetts Institute of Technology. It is a special form of double-stranded DNA that has a left-handed "zig-zag" sugar-phosphate backbone instead of the usual right-handed smooth sugar-phosphate backbone of the much more common B-form of DNA. Dr. Rich and his colleagues discovered that a crystallized form of DNA with many repeating Cytosine and Guanine base pairs (i.e., CGCGCGCG etc.) formed a left-handed coil rather than the expected right-handed coil. It has been shown that the Z-form of DNA does occur in living cells of both prokaryotes and eukaryotes. Due to its unique configuration, it could play a role in the regulation of gene expression by changing the affinity of transcription factors to DNA binding sites.

* zinc finger proteins: In many proteins, including DNA-binding proteins, there are separate structural and functional units called domains within the same polypeptide chain. Among the more than 100 DNA binding proteins in eukaryotes, the DNA binding domains fit into four structural patterns. One of these patterns, or motifs, is called the zinc finger. The other three motifs are helix-turn-helix, helix-loop-helix, and leucine zipper. The zinc-finger domain is composed of four zinc-binding amino acid residues at the base of a "finger-like looping-out" of several other amino acid residues. The zinc-binding amino acids may consist of all cysteine residues or of two cysteine and two histidine residues.

References

Byczynski, L., 1991. *Genetics: Nature's Blueprints: The Encyclopedia of Discovery and Invention.* San Diego, California: Lucent Books Inc.

Campbell, N.A., 1993. *Biology* (Third Edition). Redwood City, California: The Benjamin/Cummings Publishing Company Inc.

Connor, J.M. and Ferguson-Smith, M.A., 1993. *Essential Medical Genetics* (Fourth Edition). Oxford, England: Blackwell Publications.

Cox, M.M., Lehninger, A.L., and Nelson, D.L., 1993. *Principles of Biochemistry* (Second Edition). Worth Publishers Inc.

Edwards, Lois, 1997. *Simulations in Recombinant DNA.* Toronto, Ontario: Trifolium Books Inc. (in press).

Fritsch, E.F., Maniatis, T., and Sambrook, J., 1989. *Molecular Cloning* (Second Edition). Cold Spring Harbor, USA: Cold Spring Harbor Laboratory Press.

Galbraith, D., 1989. *Biology: Principles, Patterns and Processes.* Toronto: John Wiley and Sons Canada Ltd.

Gardner, E.J., Simmons, M.J., and Snustad, D. Peter, 1991. *Principles of Genetics* (Eighth Edition). John Wiley and Sons Inc.

Gilman, M., Watson, James D., Witkowski, J., and Zoller, M., 1992. *Recombinant DNA*. New York, N.Y.: W.H. Freeman and Company.

Grace, Eric, 1997. *Genetic Engineering and All That (tentative).* Toronto, Ontario: Trifolium Books Inc. (in press).

King, Robert C. and Stansfield, William D., 1990. *A Dictionary of Genetics* (Fourth Edition). Oxford, England: Oxford University Press Inc.

Newell, J., 1991. *Playing God?* London, England: Broadside Books Ltd.

Purves, W.K., Orians, G.H., and Heller, H.C., 1995. *Life: The Science of Biology* (Fourth Edition). Sinauer Associates, Inc. distributed by W.H. Freeman and Co.

Rifkin, J., 1983. *Algeny* (First Edition). The Viking Press. New York, N.Y.

Starr and Taggart, 1996. *Biology Concepts & Applications* (Third Edition). Wadsworth distributed by ITP-Nelson Canada.

Wessell, N.K., Hopkins, J.L., et al., 1988. *Biology.* McGraw-Hill Ryerson Limited.